OFFICIAL DISCLAIMER

The Federal Activities Report on the Bioeconomy is a product of interagency collaboration under the Biomass R&D Board and does not reflect any changes in policy. All information is based on current activities conducted by the Executive Agencies as of October 2015. The potential impacts of technical progress, pending or proposed legislation, regulations, standards, or any other changes that may have occurred after the publication date are not reflected in this report.

This page is intentionally left blank

Table of Contents

Foreword

The Biomass R&D Board

The Biomass Research and Development (R&D) Board (The Board) was created through the enactment of the Biomass Research and Development Act of 2000 "to coordinate programs within and among departments and agencies of the federal government for the purpose of promoting the use of biobased industrial products by (1) maximizing the benefits deriving from federal grants and assistance; and (2) bringing coherence to federal strategic planning."[1] The Board is co-chaired by senior officials from the U.S. Departments of Energy (DOE) and Agriculture (USDA) and currently consists of senior decision makers from the DOE, USDA, U.S. Department of Transportation (DOT), U.S. Department of the Interior (DOI), U.S. Department of Defense (DoD), U.S. Environmental Protection Agency (EPA), National Science Foundation (NSF), and the Office of Science and Technology Policy (OSTP) within the Executive Office of the President. With its diverse membership, the Board facilitates coordination among federal government agencies that affect the research, development, and deployment of biofuels and bioproducts.

Purpose of the Federal Activities Report on the Bioeconomy

A Federal Activities Report on the Bioeconomy has been prepared to emphasize the significant potential for an even stronger U.S. bioeconomy through the production and use of biofuels, bioproducts, and biopower. This report is intended to educate the public on the wide-ranging, federally funded activities that are helping to bolster the bioeconomy. Further, the report will highlight some of the critical work currently being conducted across the federal government that either supports or relates to the bioeconomy.

---

1 U.S. Congress (1999). HR 2827. 106th Congress, first sess on. http://www.gpo.gov/fdsys/pkg/B LLS-106hr2827 h/htm /B LLS-106hr2827 h.htm

BIOECONOMY
VISION

This page is intentionally left blank

Introduction

Why a Bioeconomy

"Bioeconomy" is a popular phrase used broadly in today's energy conversation. The term emphasizes the integral role of abundant, sustainable, domestic biomass in the U.S. economy.

The benefits of the bioeconomy have been well documented.[2] Biobased products support the growth of plants, trees, and vegetation, which recycle carbon (CO_2) from the atmosphere, resulting in air quality improvements when compared to fossil fuel-based products. In addition, producing and utilizing biobased fuels reduces U.S. reliance on foreign oil (energy security) and creates jobs, particularly in rural areas.

Because of these and other benefits, biobased resources are an important component of the nation's "All of the Above" energy policy, which is focused on developing a range of available energy sources to meet future demands and increase the stability and security of the nation's energy supply.[4] In the Climate Action Plan, the President recognized that "biofuels have an important role to play in increasing our energy security, fostering rural economic development, and reducing greenhouse gas emissions from the transportation sector."[5] In addition to biofuels, the federal government has emphasized the potential for biopower, bioproducts, and renewable chemicals as key components of the national bioeconomy.

> **WHAT IS A BIOECONOMY**
>
> For the purpose of this report, the bioeconomy is defined as:
>
> "The global industrial transition of sustainably utilizing renewable aquatic and terrestrial biomass resources in energy, intermediate, and final products for economic, environmental, social, and national security benefits."[3]

While the United States is a global leader in promoting the use of sustainably produced feedstocks to fuel economic activity and growth, the bioeconomy is still in its early stages. A transition is needed from a fossil-based economy to an economy that is fueled by sustainable and renewable energy, of which biomass plays a critical role. Assisted by public-private partnerships, development of new and innovative technologies in the United States is leading to renewable and drop-in fuels, biobased materials, and renewable chemicals that are replacing fossil-based products.

2 Renewab e Chem ca s & Mater a s Opportun ty Assessment - Major Job Creat on and Agr cu tura Sector Eng ne, USDA, 2014. http://www.usda.gov/oce/reports/energy/USDA_RenewChems_Jan2014.pdf; U.S. Econom c mpact of Advanced B ofue s Product on: Perspect ves to 2030, B oeconomy Research Assoc ates, 2009. https://www.b o.org/s tes/defau t/fi es/Econom c mpactAdvancedB o-fue s.pdf; B o-based Chem ca s – Va ue Added Products from B orefiner es, EA 2011. http://www. eab oenergy.com/wp-content/up-oads/2013/10/Task-42-B obased-Chem ca s-va ue-added-products-from-b orefiner es.pdf

3 An Econom c mpact Ana ys s of the U.S. B obased Products ndustry - A Report to the Congress of the Un ted States of Amer ca, USDA 2015. http://www.b opreferred.gov/B oPreferred/faces/pages/DocumentBrowser.xhtm #; Opportun t es n the Emerg ng B oeconomy, Ju y 25, 2014. www.b opreferred.gov/fi es/WhyB obased.pdf; Renewab e Chem ca s & Mater a s Opportun ty Assessment - Major Job Creat on and Agr cu tura Sector Eng ne, USDA, 2014. http://www.usda.gov/oce/reports/energy/USDA_RenewChems_Jan2014.pdf

4 https://www.wh tehouse.gov/s tes/defau t/fi es/docs/c ean_energy_record.pdf

5 https://www.wh tehouse.gov/s tes/defau t/fi es/ mage/pres dent27sc mateact onp an.pdf

Bioeconomy and the Federal Government

Bioeconomy activities have already touched on the interests of many federal agencies and offices. For example, USDA's interest stems from the development and production of feedstocks to end-use markets for those feedstocks; DOE's interest involves alternative inputs used to generate energy and co-products, technologies, and end-use markets; EPA leads development of the renewable fuel standard for biofuels, environmental approval of biotechnologies essential to the bioeconomy, regulates emissions from mobile and stationary sources of the bioeconomy, and the management of

manures, municipal solid waste and biosolids – all potential feedstocks for the bioeconomy; NSF recognizes innovative technology underpins a strong bioeconomy and increased demand for scientists and engineers; DoD and DOT view the bioeconomy from an alternative fuels and power dimension and are interested in reducing energy price variability and identifying reliable and sustainable fuel supplies for all transportation modes, as well as tactical and non-tactical assets; and DOI seeks to add value to U.S. land.

The federal government as a whole sees great potential in the nation's abundant natural resources, the capacity for new and advanced technologies, and the entrepreneurial spirit of the American people. This potential offers the ability to triple the size of today's bioeconomy by 2030. To realize the full potential of a sustainable, domestic bioeconomy, collaboration among federal agencies is needed, as are partnerships with state and local governments, universities, industries affecting the bioeconomy supply chain, and the American public. The Biomass R&D Board is a standing body that can facilitate effective implementation of this vision.

Path to the Billion Ton Bioeconomy Vision

Genesis of the Bioeconomy Vision

Since fall 2013, Board members have been planning a broad, new vision to promote the expansion of the bioeconomy. With one billion tons of biomass projected to be sustainably produced and available annually by 2030, the Board recognized the need to fully develop a Billion Ton Bioeconomy Vision.[6] The Board believes that a single, coordinated multi-department vision focused on developing and implementing a plan for utilizing this biomass for these purposes will increase economic activity, decrease reliance on foreign oil, and create market-driven demand for bioenergy and bioproducts.

The concept was discussed in parallel by the Biomass R&D Technical Advisory Committee (TAC). The TAC envisioned a new interagency effort that would help to replace more fossil carbon with renewable carbon in transportation fuels and related products. The TAC recommended that the new vision endeavor to rapidly expand emerging biofuels and bioproducts industries, targeting a potential 30% penetration of biomass carbon into the U.S. transportation market by 2030 in a sustainable and cost-effective manner to create jobs, reduce greenhouse gas impacts, and enhance national security.

6 U.S. B on-Ton Update, August, 2011. http://www1.eere.energy.gov/b oenergy/pdfs/b on_ton_update.pdf

Additional expected outcomes of a proposed Bioeconomy Vision by the TAC and Board include the following:

- Enhanced economic development by increasing direct and indirect jobs.
- A cost-effective energy supply that is synergistic with existing fossil-based energy markets.
- Enhanced economic, environment, and social sustainability.
- Improved national energy security and decreased dependence of national defense on foreign energy supplies.

The Board Operations Committee also conducted preliminary analyses to determine the feasibility of having one billion dry tons of biomass sustainably available annually within the relatively short timeframe of 15 years.[7] These bio-feedstocks would be used to produce biofuels, bioproducts, renewable chemicals, and biopower. Some preliminary analyses, led by members of the Board's Analysis Interagency Working Group, were conducted to determine the feasibility of reaching this target and harnessing some of the associated social, economic, and environmental benefits of the effort.[8]

The Goal of the Bioeconomy Vision

The goal of the Billion Ton Bioeconomy Vision (the Vision) is to develop and implement innovative approaches to remove barriers to expanding the sustainable use of America's abundant biomass resources, while maximizing economic, social, and environmental outcomes. By increasing use of renewable plant material and waste feedstocks for biofuels, bioproducts, and biopower, the Vision has the potential to stimulate job growth and economic opportunities; increase the nation's competitive advantage; support a secure, renewable energy future; and contribute to improved environmental quality. This interagency vision differs from "business-as-usual" with an intention to create a sustainability framework that considers multi-dimensional impacts and benefits from the use of biomass.

INTERAGENCY COLLABORATION

Founding Partnerships, Breaking Barriers

USDA, DOE, and EPA jointly released a progress report on the Biogas Opportunities Roadmap of 2014 The Roadmap identifies voluntary actions that can be taken to reduce methane emissions through the use of biogas systems It outlines strategies to overcome barriers limiting further expansion and development of a robust biogas industry in the United States

For more information on this activity and others like it, see Appendix II

7 The B on-Ton Update assumes that the geograph c range where energy crop product on can occur s m ted to areas where product on s under ra n-fed cond t ons, w thout the use of supp ementa rr gat on.

8 Ana ys s nteragency Work ng Group was estab shed as part of the 2012 Nat ona B ofue s Act on P an. The focus of the group s to dent fy and address ana ys s needs n b ofue s and b oproducts R&D. The WG s co- ed by the Departments of Agr cu ture and Energy and has representat ves from these departments, EPA, and DOT. Th s Group, a ong w th added members from nat ona aborator es, prov ded nput and recommendat ons to the n t a B oeconomy V s on ana ys s, as we as a forma rev ew of the assumpt ons, data, references, and ca cu at ons. There were more than 10 members nc ud ng sc ent sts and program managers w th expert se n agr cu tura, forestry, mode ng, stat st cs, GHG em ss ons, econom cs, transport systems, and many other areas.

Biomass already plays a significant role in the Administration's "All of the Above" energy strategy. The Vision is to take a systems approach to sustainably reach the full potential of biomass-derived products as a way of expanding our nation's economy.

This effort will increase the sustainable production of biomass feedstocks and capture of usable wastes; development of innovative and more efficient technologies to transform renewable carbon to intermediates and products; construction of more biorefineries and manufacturing facilities; and expansion of the market for biofuels, biochemicals, biopower, and other biomass-derived products. As a result, the bioeconomy will provide multiple economic, environmental, and social benefits to the nation.

The Vision requires new science and technologies from universities and laboratories; industry and manufacturing development; engagement with financial institutions; education and job training; and additional producers, contractors, and specialty personnel. Expanding the economic sector to the extent envisioned in such a short timeframe will require strong support, commitment, and involvement of many sectors and groups. The Board's Operations Committee will coordinate engagement and outreach efforts to achieve commitment and investment among these external stakeholders. The Committee will collaborate with bioeconomy partners to develop a strategic framework and plan of action for reaching target groups across various sectors.

The Role of the Biomass R&D Board

The purpose of the Bioeconomy Vision is to expand the sustainable production and use of biomass. The Board will have the primary responsibility for providing overall leadership and coordination of this vision. The Vision will coordinate and enhance federal efforts, as well as facilitate collaboration between the government and its stakeholders.

There is already a high level of commitment and coordination among many federal agencies, as well as a fairly comprehensive approach to remove barriers that may preclude the commercialization and expansion of the bioeconomy. To maximize the potential of the U.S. bioeconomy, the Vision seeks to further align and coordinate agency missions, and fully engage with the public to achieve success. By involving other agencies and stakeholders as part of a collaborative effort, the Vision will strengthen this commitment and coordination, while broadening the effort to assess impacts and benefits, and prioritize the design and efficient scale-up of production and utilization of the best biomass production and utilization pathways.

Federal departments, agencies, offices, and programs will provide the following:[9]

- Investments in research, development, and demonstration (RD&D), as well as an objective assessment of life-cycle performance of technology and system choices looking at RD&D, production, conversion, delivery, and use.

9 As a separate body estab shed to nform and adv se the Board, the B omass R&D Techn ca Adv sory Comm ttee, an ndependent rotat ng adv - sory group of about 30 b oenergy stakeho ders, w be encouraged to prov de recommendat ons on the B oeconomy V s on to the Board.

- Workforce development, training, public outreach, extension/technology transfer.
- An understanding of the policies of all agencies to successfully implement regulations that enable and overcome challenges to the sustainable expansion of the bioeconomy.
- New and existing mechanisms to leverage public/private relationships and resources.
- DOE national labs, USDA, and other federal research facilities that

will apply their technical expertise and provide technologies to further break down technical barriers, and decision support tools for the advancement of a sustainable bioenergy industry.

Overview of the Billion Ton Bioeconomy Vision

Scope of the Bioeconomy Vision

The Vision includes assessing benefits and impacts, and evaluating approaches for sustainably expanding the production of biomass-based fuels, power, and products along the entire supply chain (from the production of feedstocks to their conversion and end use). These efforts will require analyses, financial mechanisms, contract services, research, development, and deployment for feedstocks, conversion technologies, manufacturing, infrastructures, policies and regulations, and market development of products.

Expanding the bioeconomy in a sustainable manner will increase energy diversity and long-term security. It will provide additional economic, environmental, and social benefits, such as reduced greenhouse gas emissions, job growth, and responsible management of diverse sources of biomass and waste materials. Efforts will result in a greener, stronger nation with diverse, new economic sectors that enhance U.S. competitiveness.

INTERAGENCY COLLABORATION

For more information on this activity and others like it, see Appendix II

A More Sustainable Future Together

DOE and USDA engage with the Global Bioenergy Partnership to exchange information and findings on sustainability and greenhouse gas emissions of bioenergy The Global Bioenergy Partnership (GBEP) was established in 2006 by a mandate from the G8 + 5 in 2005 and is the venue for the U S government to discuss sustainability of bioenergy on an international level Partners currently include 23 countries and 7 United Nations organizations, with Brazil and Italy as co-chairs

The following are examples of existing and proposed Bioeconomy Vision approaches:

- The Vision will be developed with participation from federal agencies that are not currently involved with the Biomass R&D Board, thereby expanding the commitment of the federal government to identify and develop sustainable pathways in support of the bioeconomy.

- The Vision will proportionately expand its focus on biofuels to include other uses of biomass to derive increased value and benefits.

- A national framework will be developed to coordinate federal and external efforts to advance the bioeconomy.

- Emphasis will be placed on achieving an overall national bioeconomy goal; specific milestones and deadlines for that goal will be established.

Expected Benefits

Environmental Benefits

An expanded bioeconomy must help mitigate, and not exacerbate, challenges associated with resources, environment, and public health. The Vision is to potentially help address many of today's pressing environmental challenges: The preservation and/or enhancement of ecosystem services and biodiversity; improved materials, water, and energy conservation; shift from non-renewable to renewable feedstock availability, selection, and use in industrial and consumer product manufacture; capture and recovery of non-renewable and renewable materials from wastes; improved air quality; and improved public health. Environmental assessments can identify pathways that would result in  environmental gains when accounting for their full life cycle. Some environmental benefits are highlighted below:

- **Reduced Greenhouse Gas Emissions** - Greenhouse gas emission footprints associated with biomass-derived products are often more favorable than their fossil-fuel derived counterparts. Preliminary analyses show that a Billion Ton Bioeconomy could potentially reduce GHG emissions by over 400 million tons annually of CO_2e, which is equivalent to an approximate 8% reduction in current U.S. emissions.

- **Optimizing Land Use** – The sustainable production of biomass makes use of multi-functional landscapes that improve ecosystem services such as soil health, water quality and availability, and accessibility of other natural resources for food, fuel, and fiber production. For example, the development of non-food crops capable of thriving on marginal lands, which require less inputs for production, often helps reduce nutrient run-off, soil erosion, and water use. Also, aquatic biomass developed on coastal and marine landscapes can provide similar benefits and expand the areal production of biomass for the bioeconomy.

- **Higher Purpose Use of Biobased Materials** - Opportunities exist to develop and utilize diverse agricultural, algal, and biogenic waste (e.g. manures, biosolids, food waste, and municipal solid waste) feedstocks that, historically, have presented environmental and economic challenges. Using diverse sources of renewable biobased materials as feedstocks often reduces greenhouse gas, eliminates management costs, reduces water quality degradation, and provides revenue streams and renewable energy.

- **Capture, Re-use, and Market Value of Non-renewable Resources and Pollutants from Biogenic Wastes** – Innovative technologies can extract non-renewable resources (such as phosphorus from manures and biosolids) and transform them from costly pollutants into marketable products.

Economic Benefits

The Analysis Interagency Working Group of the Biomass R&D Board estimates that the bioeconomy currently contributes approximately $50 billion and over a quarter million jobs to the U.S. economy. The government's Billion Ton Bioeconomy Vision has established goals that would considerably increase U.S. economic production and job growth. Effects of a significantly larger bioeconomy would ripple across all U.S. economic sectors, creating new jobs in biomass production and logistics, facility operation and quality control, and distribution, as well as R&D, finance, supplies, and services that deal with the complexities of building and running new biorefineries. If targets are met, the bioeconomy could achieve the following milestones:

- **National Revenues**: The total direct revenue could reach approximately $250 billion annually, with a total economic impact of $660 billion each year including indirect economic outputs from the bioeconomy.
- **Job Growth**: The cumulative job benefit could be over a million new positions that cannot be outsourced from the United States. A strong bioeconomy will boost job growth in various sectors, especially in the high-paying technology field. Such progress would have positive effects throughout much of the agriculture, aquaculture, forestry, and service sectors of the economy.
- **Rural Development**: Sustainably utilizing our land resources, algae, and waste systems would provide economic development opportunities for both rural and urban America. Additionally, rural communities and economies would have the potential to benefit significantly as a result of the Bioeconomy Vision.

National Security Benefits

The volatility of oil prices and supply is a major concern to a nation still reliant on oil for the majority of its transportation needs. Biofuels are less susceptible to global events and can help bolster the U.S. renewable energy markets. Additionally, an expanded bioeconomy will contribute to the U.S. maintaining a competitive advantage in a global market. Specific benefits include the following:

- **Price Volatility**: The Billion Ton Bioeconomy will create more biobased energy options and provide flexibility to address environmental challenges and future volatility in global fuel prices.
- **Infrastructure Compatibility**: The bioeconomy will support development of a responsive, reliable, and efficient transport and distribution infrastructure that can safely deliver biofuel products to their end-use locations.

INTERAGENCY COLLABORATION

Scaling up for Success

To supply the U S Navy with a home grown, secure fuel supply, USDA, DOE and the Navy have been working together since 2011 to help commercialize new fuels that meet military specifications for jet fuel and diesel These domestically produced renewable biofuels will help the U S military increase the nation s energy security, reduce greenhouse gas emissions, and create jobs in America

For more information on this activity and others like it, see Appendix II

Challenge Areas

There is general agreement that expanding the bioeconomy would offer significant benefits; however, some major hurdles still exist. They include: (1) a lack of available, cost-competitive, biomass-derived products as compared to fossil resource (petroleum, natural gas, coal) products, (2) concerns over environmental issues associated with growing biomass and mitigating/reducing negative impacts, (3) reducing risks to warrant investment in biomass production systems, conversion facilities, and end-use infrastructure, (4) limited availability of land and resources (e.g., water, fertilizer, labor, etc.) to produce one billion tons of biomass, (5) current inability to transport and dispense larger quantities of fuels, and (6) the need for large capital expenditures in

a risk-averse financial environment. The goals of the Vision can only be accomplished when these challenges have been considered and mitigated. At the same time, it is critical that U.S. food, feed, and fiber production is sustained and enhanced.

There are four main barriers that restrict our ability to achieve Billion Ton Bioeconomy goals:

- **First, sustainably producing and accessing adequate, affordable feedstocks** - Biomass as the source of low-cost renewable carbon feedstock for conversion to fuels, power, and products add significant complexity to the agricultural and forestry industries for the production of biomass and to the waste management industries for the recovery of organic materials

- **Second, developing and applying innovative, cost-competitive conversion technologies** - Conversion technologies for production of fuel from cellulosic feedstocks suffer from high energy requirements and low productivity (yield, selectivity, and rate of production), as well as high capital expenditures per gallon, which results in conversion technologies that are unable to achieve reinvestment economics Solid fuels combustion systems are relatively commercial state-of-the-art technologies, but are still significantly inefficient conversion processes for electricity and thermal energy production Current conversion technology cannot economical maximize value from by- and co-products Using and blending waste resources into conversion system are still limited by non-robust conversion technologies

- **Third, optimizing distribution infrastructure across the nation to allow movement of biomass and subsequent derivatives across the entire supply chain** - The lack of an updated distribution infrastructure and other market incentives, especially for fuels, directly impacts their use by consumers Developing functional, compatible, and competitive distribution systems for end use, while expanding existing infrastructure (i e , feedstocks, distribution, end-use, and delivery) is critical to the success of the Bioeconomy Vision

- **Fourth, educating the consumer** – Education of the positive aspects of biomass, re-use of waste streams, and production of biofuels and bioproducts

Proposed Objectives of the Bioeconomy Vision:

- By leveraging existing expertise across the various agencies, this new vision could propel the United States to develop the leading bioeconomy in the world. The objectives of the Vision can be categorized into five distinct areas. A table has been provided to discuss the main themes of the Bioeconomy Vision.

Use an integrated systems approach 1	Holistic, integrated supply chains that overcome barriers and reduce financial, environmental, and market risks Objectives are • Develop and deploy sustainable biomass systems that improve cost competitiveness and mitigate economic risk, while maintaining or enhancing environmental quality throughout their life cycle • Utilize financial data and business models to reduce risks in commercialization in order to satisfy needs along the entire supply chain • Utilize models and data across the federal government to understand and quantify tradeoffs and synergies to optimize the economic, environmental, and social benefits of the bioeconomy to minimize adverse impacts
Provide the science and the technology 2	Science provides solutions to barriers and innovative technologies help drive the bioeconomy Objectives are • Make significant advances in transformational science and technology, throughout the supply chain, that are resource efficient, cost-effective, and environmentally neutral or enhancing • Efficiently integrate and validate engineering, environmental, and economic data that are associated with biomass production, conversion to bioeconomy products, and use • Adapt current infrastructure and design and develop new machinery for efficiency • Integrate biophysical, environmental, and social models in systems development
Public and private collaboration to overcome barriers and accelerate deployment 3	Expansion of the bioeconomy requires public and private collaboration across the entire sector federal, state, and local government, tribes, business and industry, academia, producers, landowners, workers, and many others Objectives are • Identify and understand roles of the various factions and develop optional mechanisms for information sharing, consensus building, collaboration, and cooperation • Work with producers (farmers, landowners/managers, waste managers) and communities to encourage/incentivize cellulosic and waste material biomass system adoption while understanding and mitigating risk

Develop a workforce for the future bioeconomy	Millions of additional workers are needed in agricultural, aquatic, and forestal production, biorefineries R&D, transportation, manufacturing, and various allied fields Objectives are
4	• Accelerate the emergence of trained professionals in all functions needed to establish a sustainable bioeconomy, including project design and production, transport, market penetration and use, specialized financial management, and regulatory approval • Develop specific educational programs for professionals and technical students at various levels of higher education • Provide career pathways information and activities for high school students
Understand and inform policy	Policies seek to improve economic, environmental, and social outcomes, and drive direction and funding of new scientific endeavors and program implementation to accelerate the bioeconomy Objectives are
5	• Analyze and understand the impact of national policy on the bioeconomy • Inform local and state governments, industry, and other stakeholders about policies and their impact • Integrate economic and environmental policy factors to reduce financial risks to investors, producers, and manufacturers, and to protect environmental quality across the supply chain

Federal Agencies' Current Work in the Bioeconomy

Executive Agencies

As stated previously in this report, the bioeconomy involves many different agencies and mission areas within the federal government. The national bioeconomy is wide-reaching and multi-faceted, and relates to a range of different offices within many executive agencies, even beyond those represented on the Biomass R&D Board. Appendix I covers all Board agencies, and details specific offices and programs involved in the current bioeconomy in support of their specific mission areas (note that this is not an exhaustive listing for the entire federal government). The following chart presents a snapshot of the synergies and roles for each agency across the vast bioeconomy supply chain. For details on how each agency is currently supporting specific aspects of the bioeconomy, please see Appendix I.

FEEDSTOCK SUPPLY

BIOMASS CONVERSION

BIOENERGY DISTRIBUTION

BIOENERGY END USE

Agency	Feedstock Supply	Biomass Conversion	Bioenergy Distribution	Bioenergy End Use
DOE	●●● ●	●●●● ●	● ● ●	●●● ●●
USDA	●● ●●	●● ●	●● ●●	● ●●
DOT	● ●●	● ●	●● ●●	●● ●●
EPA	●● ●●	●●● ●	●●● ●	●● ●
DOI	● ●	●		
NSF	●● ●	●●● ●	●	
DoD		● ●	● ●●	● ●●

Legend:

● Use an integrated systems approach
● Provide the science and the technology
● Public and private collaboration to overcome barriers and accelerate deployment
● Develop a workforce for the future bioeconomy
● Understand and inform policy

Interagency Collaboration

As evident throughout this report, there is already significant collaboration between agencies to take advantage of the wide-ranging expertise across the government. Interagency activities take many forms and seek to confront varied challenges facing the bioeconomy today. These collaborative projects offer the chance to fully leverage federal government expertise, ensuring that activities produce valuable projects and achievements for the public.

Further, interagency projects offer some of the best opportunities for stakeholder organizations to work directly with multiple agencies on specific aspects of the bioeconomy. For more examples of interagency projects currently underway, please see Appendix II.

Looking Ahead

Bioeconomy Federal Strategy Workshop

In May 2015, the Board convened a workshop of federal agencies to discuss options for expanding and developing the bioeconomy. A significant component of this activity involved sharing information on existing agency programs and activities that intersect with or are directly focused on the bioeconomy. The federal government has numerous efforts underway to help expand the bioeconomy, and brief summaries of those activities are included in this report.

The workshop included many productive discussions given the extensive knowledge, experience, opinions, and insights across the agencies. Participants addressed issues concerning specific aspects of the supply chain, while also considering the sustainable expansion of the bioeconomy from a holistic perspective. Exchanges among the participants were enlightening, concluding that the effort would be challenging, but not insurmountable. Some key discussion points involved the following:

- Addressing technical and sociological challenges to using genetically modified organisms (GMOs) in developing sustainable feedstocks.
- Identifying the key role algae plays for the future of bioproducts and biochemicals. Sharing a common understanding of sustainability and achievement metrics, and discussing paths for practical implementation.
- Realizing that biomass production and use must be within the context of sustainability, resource availability, social concerns, and public perceptions.
- Being aware that, if done correctly, an expanded bioeconomy could significantly impact America's landscape with opportunities to improve ecological functions and ensure the mitigation of negative impacts of increased land use.
- Expansion of the bioeconomy will happen in the marketplace, and the market must simultaneously meet society's need for food, fiber, and forage, as well as energy and products.

Overall, the workshop helped federal agencies identify processes for working together and with their many stakeholders. The following specific efforts were established at the meeting and are being addressed:

- Ensure consistent federal messaging
- Increase federal agency coordination
- Recognize opportunities for improvements across the supply chain
- Integrate diverse national goals and objectives linked to the sustainable bioeconomy
- Coordinate stakeholder interests and actions in both the public and private sectors

Notably, the workshop became the bridge to building a national effort within the federal government to work together to expand the bioeconomy. Although many barriers and opportunities were identified and discussed, issues were not resolved at the meeting. The workshop was not intended to establish processes or undertake detailed planning. The workshop was the cornerstone for identifying roles and issues and building a vision that would conceivably bring together many federal agencies, stakeholders, and the American public.

Building a National Coalition

The work has already started on building a national federal government coalition with the goal to expand the bioeconomy. Agencies have distinctive missions that relate to the bioeconomy that are currently being accomplished through various approaches and programs, authorized by variousstatutes. This coordinated effort will leverage these approaches and programs, as appropriate.

As described previously, many of these agencies have been coordinating similar efforts for over a decade through the Biomass R&D Board. The Board has the lead role in building the coalition while working through the agency's executive leaders, managers, and staff. However, this activity will not only involve agencies, but will be conducted in cooperation with many stakeholders and with the public. This report is the first step in building the coalition—the next step is engaging the public, beginning with agency stakeholders. From this collaborative engagement, an implementation plan will be developed to help direct the Vision.

The federal government's goal to expand biomass resources for economic, environmental, social, and energy security benefits will be facilitated by collaborative educational and engagement efforts. The global transition to full utilization of biopower, bioenergy, and biobased products is being driven by a wide array of groups. Reaching out to diverse stakeholders (non-governmental organizations; international governments and organizations; environmental and industry groups; manufacturers; and other members of the supply chain) to obtain multiple perspectives, share key data, identify opportunities, and understand motivations, business cases, and realities will help achieve the bioeconomy goal and its expected benefits.

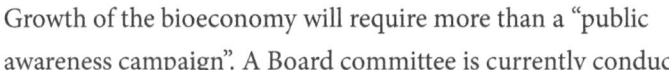

Growth of the bioeconomy will require more than a "public awareness campaign". A Board committee is currently conducting efforts to identify and understand more about the various bioeconomy stakeholders. This effort will help determine the Board's future education and communications activities to address current barriers to growing the bioeconomy. These activities will include meaningful communication across the U.S. government (developing and deploying consistent federal messaging); stakeholder and public workshops (listening sessions to gather information about the Bioeconomy Vision and the role of education); and the potential establishment of a formal coalition to provide leadership and oversight of the Vision's strategy and practice. The information gathered from this direct engagement with the public will result in an overall plan for the Vision, to be detailed in an Integrated Bioeconomy Implementation Plan, to be released in 2016.

Appendix I: Federal Agency Activities

This Appendix includes details on the missions, activities, and linkages to the U S Bioeconomy within the Executive Agencies participating on the Biomass R&D Board Each agency brings their own mission and expertise to the issues surrounding creating a bioenergy industry Numerous sub-agencies, divisions, and program offices are included below, however due to the broad nature of the bioeconomy, there are more offices that play a part in this sector that are not listed Additionally, information included in this appendix is only a snapshot of the work being done by each office For additional details on these offices and agencies, please visit the website listed

DEPARTMENT OF ENERGY

www.energy.gov

Bioenergy Technologies Office

The Bioenergy Technologies Office seeks to develop and transform our renewable biomass resources into commercially viable, high-performance biofuels, bioproducts, and biopower through targeted research, development, and demonstration supported through public and private partnerships.

This Office conducts R&D, demonstration, and market transformation activities through an integrated supply chain approach addressing supply, conversion, distribution, and end use. Cross-cutting areas—sustainability, strategic analysis, and strategic communications—develop and disseminate a body of knowledge and tools related to the economic, environmental, and social dimensions of advanced bioenergy. The process is managed through its Multi-Year Program Plan (MYPP).[1] Program of work includes feedstock supply assessment, logistics systems improvements, lower cost conversion technologies, and market development. Signature elements are the periodic national assessment, advanced and depot logistics, validated conversion pathways, feedstock/conversion interfaces and integration, market development, technoeconomic analyses, and sustainability analysis and research. Work is done through national laboratories, competitive R&D grants, consortia, and partnerships. The Office collaborates with other DOE Offices and departments.

The Bioenergy Technologies Office is dedicated fully developing the bioeconomy, and establishes partnerships with key public and private stakeholders to develop and demonstrate technologies for producing cost-competitive advanced biofuels from non-food biomass resources, including cellulosic biomass, algae, and wet waste (e.g. biosolids). The Office pursues innovations for and testing of crucial bioenergy technologies including pilot- and demo-scale integrated biorefineries.

Vehicle Technologies Office

The Vehicle Technologies Office supports research and development that will lead to new technologies that reduce our nation's dependence on imported oil, further decrease vehicle emissions, and serve as a bridge from today's conventional powertrains and fuels to tomorrow's hydrogen-powered hybrid fuel cell vehicles. The Vehicle Technologies Office also supports implementation programs that help to transition alternative fuels and vehicles into the marketplace, as well as collegiate educational activities to help encourage engineering and science students to pursue careers in the transportation sector.

The Office supports research to improve how vehicles will use alternative fuels in the future. Work includes determining the impact of biofuels' properties on engines' efficiency, performance, and emissions. Activities include examining ways to increase alternative fuel vehicles' fuel economy, investigating the potential effects of upcoming

1 http://www.energy.gov/eere/bioenergy/downloads/bioenergy-technologies-office-multi-year-program-plan-july-2014-update

blends, and improving the quality of current and future biofuel blends, especially biodiesel and E85. VTO's Clean Cities Program develops public/private partnerships to promote alternative fuels, vehicles, and infrastructure.

Fuel Cell Technologies Office

The Fuel Cell Technologies Office conducts comprehensive efforts to overcome the technological, economic, and institutional barriers to the widespread commercialization of hydrogen and fuel cells. The office's efforts will help secure U.S. leadership in clean energy technologies and advance U.S. economic competitiveness and scientific innovation.

This is the lead federal agency for applied research and development (R&D) of cutting edge hydrogen and fuel cell technologies. The overall challenge to hydrogen production is cost reduction which includes the use of biomass. DOE is working closely with its national laboratories, universities, and industry partners to overcome critical technical barriers to fuel cell commercialization.

Advanced Manufacturing Office

The Advanced Manufacturing Office partners with industry, small business, universities, and other stakeholders to identify and invest in emerging technologies with the potential to create high-quality domestic manufacturing jobs and enhance the global competitiveness of the United States.

Efforts leverage state, utility, and local resources to help manufacturers save energy, reduce climate and environmental impacts, enhance workforce development, and improve national energy security and competitiveness throughout the supply chain. The Office invests in emerging technologies that include the use of biomass for fuels, chemicals, materials, heat, and electricity. A specific example is developing polyacrylonitrile for the manufacture of carbon fiber.

Federal Energy Management Program Office

The Federal Energy Management Program (FEMP) works with key individuals to accomplish energy change within organizations. It brings expertise from all levels of project and policy implementation to enable federal agencies to meet energy-related goals and to provide energy leadership.

The Program provides technical assistance to reduce the energy intensity of federal facilities. FEMP also helps federal agencies with funding mechanisms for their projects, such as energy-saving performance contracts (ESPCs). Such projects could include the use of biomass. FEMP works with the federal fleet to increase the use of biopower, renewable and alternative fuels, and flexible-fuel vehicles.

Office of Fossil Energy

The Office of Fossil Energy (FE) plays a key role in helping the United States meet its continually growing need for secure, reasonably priced and environmentally sound fossil energy supplies. Put simply, FE's primary mission is to ensure the nation can continue to rely on traditional resources for clean, secure and affordable energy while enhancing environmental protection.

The Office's primary mission is to ensure the nation can continue to rely on traditional resources for clean, secure and affordable energy while enhancing environmental protection. The Office includes the National Energy Technology Laboratory. There are opportunities to integrate biopower applications into utility-scale electric generation and to use biomass and natural gas synergistically to maximize outputs.

Advanced Research Projects Agency-Energy

The Advanced Research Projects Agency-Energy (ARPA-E) advances high-potential, high-impact energy technologies that are too early for private-sector investment. ARPA-E awardees are unique because they are developing entirely new ways to generate, store, and use energy.

The focus of ARPA-E is on transformational energy projects that can be meaningfully advanced with a small investment over a defined period of time. ARPA-E empowers America's energy researchers with funding, technical assistance, and market readiness. An example is Plants Engineered to Replace Oil, PETRO, with its aim to develop non-food crops that directly produce transportation fuels. The PETRO activity conducts applied research and development to generate novel crop feedstocks that are directly a production of fuel molecules in planta, and improved photosynthetic activity. The production of new crop feedstocks is critical for advanced biofuels, bioproducts, and biochemicals, and the PETRO project will help develop the technology to expand the national bioeconomy. Another activity is the Transportation Energy Resources from Renewable Agriculture (TERRA) program, which conducts applied research and development of improved high throughput breeding tools (phenotyping/genotyping) to increase yields of energy crops. The expedited development of improved varieties of bioenergy feedstocks is what links TERRA to the Bioeconomy.

Office of Science: Office of Basic Energy Sciences

The Office of Basic Energy Sciences (BES) supports fundamental research to understand, predict, and ultimately control matter and energy at the electronic, atomic, and molecular levels in order to provide the foundations for new energy technologies and to support DOE missions in energy, environment, and national security. BES also plans, constructs, and operates major scientific user facilities to serve researchers from universities, national laboratories, and private institutions.

Within BES, the Physical Biosciences and Photosynthetic Systems programs support basic research to understand energy capture, conversion, and storage in plants and microbes.[2] The Physical Biosciences program combines experimental and computational tools from the physical sciences with biochemistry and molecular biology to understand the conversion and storage of energy in living systems. Supported research includes characterization of the biochemical and biophysical principles determining the architecture and biosynthesis of biopolymers and the plant cell wall and studies of redox and active site protein chemistry that can provide a basis for development of highly selective and efficient bio-inspired catalysts. The Photosynthetic Systems program brings together biology, biochemistry, chemistry, and biophysics to uncover the fundamental science of biological capture of sunlight and its conversion to and storage as chemical energy. This research provides a scientific knowledge base that can inspire artificial photosynthesis and enable new strategies for more efficient generation of biomass as a renewal energy source. The Catalysis Science program develops fundamental understanding of the scientific principles enabling rational catalyst design and chemical transformation control.[3] Supported research includes basic studies of catalysis relevant to the conversion and use of renewable energy resources and the creation of advanced chemicals. The DOE Energy Frontier Research Centers Program[4] aims to accelerate transformative discovery to understand and manipulate matter

2 http://sc ence.energy.gov/bes/csgb/research-areas/photochem stry-and-b ochem stry/

3 http://sc ence.energy.gov/bes/csgb/research-areas/cata ys s-sc ence/

4 http://sc ence.energy.gov/bes/efrc/

on the atomic and molecular scales and address key scientific challenges for energy technologies. Located across the United States, the 32 multi-investigator, multidisciplinary Energy Frontier Research Centers bring together leading scientists to tackle the toughest scientific challenges preventing advances in energy technologies. Their efforts are laying the scientific groundwork for fundamental advances in solar energy, biofuels, transportation, energy efficiency, and carbon capture and sequestration.

BES research programs provide the fundamental knowledge base of chemical and biochemical processes and properties that underpins development of new methods and technologies for efficient and sustainable production of biofuels and bioproducts.

Office of Science: Office of Biological and Environmental Research

The Office of Biological and Environmental Research (BER) supports fundamental research aimed at expanding foundational knowledge of biological systems and enabling the development of secure and sustainable bioenergy solutions.

This Office supports basic research programs relevant to DOE missions in energy and the environment. To enable the development of sustainable bioenergy and bioproducts, BER focuses on genome driven systems biology approaches that advance understanding of plants, microbes, integrated communities, and their interactions with environmental variables. This foundational understanding is used to further the development of more sophisticated biosystems design capabilities applicable to improving biomass feedstock plants, microbes applicable to biomass deconstruction and conversion to biofuels/bioproducts, and photosynthetic microorganisms. BER research also examines the intersection between plants, their associated microbiomes, and their ecosystems in order to understand responses to changing environmental variables and identify factors critical to sustainable biomass production. BER science programs are conducted via research funding at DOE National Laboratories, competitive grants to academic institutions, cooperative agreements, and support of national scientific user facilities.

BER research programs advance fundamental understanding of biosystems of plants, microbes, and biological communities and enable the development of more sophisticated biodesign approaches for sustainable production of biofuels and bioproducts.

Loan Programs Office

The Loan Programs Office (LPO) accelerates the domestic commercial deployment of innovative and advanced clean energy technologies at a scale sufficient to contribute meaningfully to the achievement of our national clean energy objectives—including job creation; reducing dependency on foreign oil; improving our environmental legacy; and enhancing American competitiveness in the global economy of the 21st century.

LPO enables DOE to work with private companies and lenders to mitigate the financing risks associated with clean energy projects to encourage the development of clean energy on a broader and much-needed scale. The Loan Programs consist of three separate programs managed by two offices, the Loan Guarantee Program Office (LGP) and the Advanced Technology Vehicles Manufacturing Loan Program Office. LPO originates, guarantees, and monitors loans to support clean energy projects, including large scale biorefineries, through these programs.

DEPARTMENT OF AGRICULTURE

www.usda.gov

Agricultural Marketing Service

The Agricultural Marketing Service (AMS) administers programs that create domestic and international marketing opportunities for U.S. producers of food, fiber, and specialty crops.

The AMS Transportation Services Division (TSD) serves as the definitive source for economic analysis of agricultural transportation from farm to market. TSD experts support domestic and international agribusinesses by providing market reports, economic analysis, transportation disruption reports, technical assistance, and outreach to various industry stakeholders. Tracking developments in truck, rail, barge, and ocean transportation, TSD provides information and analysis on the four major modes of moving food from farm to table, port to market. On every first Wednesday of the month, and with data from public documents provided by the railroads, TSD updates rail tariff rate information for Ethanol[5] and Dried Distillers Grains with Solubles (DDGS).[6] Information provided through the updates is only for reference. Individual railroads should be consulted prior to conducting any business transactions.

AMS biofuels market reports and market intelligence reports apprise stakeholder of prices and volumes of select feedstocks, fuels, and transport metrics.

Agricultural Research Service

The Agricultural Research Service (ARS) is the chief scientific in-house research agency of the USDA which conducts research to develop and transfer solutions to agricultural problems of high national priority and provide information access and dissemination to:

a. ensure high-quality, safe food, and other agricultural products

b. assess the nutritional needs of Americans

c. sustain a competitive agricultural economy

d. enhance the natural resource base and the environment, and

e. provide economic opportunities for rural citizens, communities, and society as a whole.

ARS priorities[7] are in two parts. Part One includes the U.S. regional based feedstocks production systems with a focus on sustainability and economic impact. ARS conducts advanced biofuels research at the USDA Regional Biomass Research Centers[7a] through biophysical and economic models validated for regional conditions to incorporate energy production into existing agricultural systems. Genetic improvement work is primarily directed at peren-

5 http://www.ams.usda.gov/serv ces/transportat on-ana ys s/datasets/ethano

6 http://www.ams.usda.gov/serv ces/transportat on-ana ys s/datasets/ddgs

7 http://www.ars.usda.gov/SP2UserF es/Program/213/B oenergyResearchStrategy2010.pdf

7a http://www.ars.usda.gov/research/programs/programs.htm?np code 213&doc d 24694

nial grasses including switchgrass and Napier grass, non-food biomass sorghum including sweet sorghum, energy cane, and lipid seed crops including industrial rapeseed and field pennycress. ARS also is developing technology to sustainably harvest crop residues including corn stover, and utilize livestock manure, as well as develop inventories for removing invasive tree species and restoring western rangelands.

Part Two includes developing conversion technologies at its four ARS utilization centers (Albany, CA; New Orleans, LA; Peoria, IL; Wyndmoor, PA) to enable sustainable commercial production of biofuels by the agricultural sector in ways that enhance our natural resources without disrupting existing food, feed, and fiber markets. Research will optimize both the production of plant feedstocks and the biorefining of agricultural materials to bioenergy and value-added (biobased) coproducts. This research will strengthen rural economies, provide increased supplies of renewable transportation fuel, enhance energy security, and improve the U.S. balance of trade. ARS is committed to diversifying rural economies and employment through new biobased technologies and commercial coproducts.

Animal and Plant Health Inspection Service

The Animal and Plant Health Inspection Service (APHIS) protects and promotes U.S. agricultural health, regulates genetically engineered organisms, administers the Animal Welfare Act and carries out wildlife damage management activities.

Biotechnology Regulatory Services (BRS) implements the APHIS regulations for certain genetically engineered (GE) organisms that may pose a risk to plant health. APHIS coordinates these responsibilities along with the other designated federal agencies as part of the Federal Coordinated Framework for the Regulation of Biotechnology. APHIS' Plant Protection and Quarantine program vigilantly protects agriculture and the environment against pest and disease threats to ensure a diverse natural ecosystem and an abundant and healthy food supply for all Americans. An invasive pest is a non-native species whose introduction into the country can cause damage to the economy, natural resources, or human health.

Climate Change Program Office

Climate-smart agriculture (CSA) promotes "production systems that sustainably increase productivity [and] resilience (adaptation), reduc[ing]/remov[ing] greenhouse gases (mitigation), and enhancing achievement of national food security and development goals" (FAO 2015). The U.S. is a member of the Global Alliance for Climate-smart Agriculture (GACSA), a partnership of countries and organizations committed to sustainability, resilience, and greenhouse gas mitigation in agricultural systems in a changing climate. Federal and private research produces a diversity of practices, technologies, agricultural varieties and tools to assist farmers and ranchers in mitigating and adapting to climate change. In addition to critical research and research support, USDA provides data standardization, knowledge sharing, and collaborative platforms; monitoring and decision support tools; outreach and extension activities; and a far-reaching organizational infrastructure that allows the most effective transmittal of up-to-date information between scientists and producers, and among producers themselves, leading to favorable and regionally-specific outcomes.

The USDA Climate Change Program Office also focuses on assessing the lifecycle GHG emissions of Corn-Based Ethanol. EPA's 2010 Regulatory Impact Analysis (RIA) of the revised Renewable Fuel Standard (RFS2) included a lifecycle analysis (LCA) of the GHG emissions associated with corn-based ethanol. The LCA concluded that substituting (on an energy equivalent basis) corn-based ethanol for gasoline in transportation fuels would result in a reduction in CO_2 emissions of 20-21 percent by 2022. EPA's LCA was based on projected emissions pathways through 2022 for 11 distinct GHG source categories associated with corn-ethanol. Since 2010, new scientific studies, data, and industry trends have emerged that show the emission paths of several key categories have not developed as projected, and, the actual paths strongly suggest the GHG profile of corn ethanol is significantly better than the RIA LCA concluded. For example, the largest source category in the RIA LCA is projected emissions from indirect land use change - due to future clearing of tropical forest (mainly in Brazil) to expand commodity production. Actual data now show that in the period U.S. corn ethanol increased from 3.0 billion gallons to just under 14 billion gallons, deforestation in Brazil's Amazon dropped from 10,200 square mile to just under 2,400 square miles. With renewable fuels and GHG mitigation becoming higher policy priorities, USDA's Climate Change Program, Office has contracted with ICF International to do a retrospective analysis of the RIA LCA including a review of new information that has become available since 2010 related to each emissions category, quantification of new emissions estimates for each source category, development of an updated LCA reflecting the current GHG profile of corn ethanol.

Departmental Management

Departmental Management (DM) provides management leadership to ensure that USDA administrative programs, policies, advice and counsel meet the needs of USDA program organizations, consistent with laws and mandates; and provide safe and efficient facilities and services to customers.

DM's Environmental Management Division (in the Office of Procurement and Property Management) develops policy and implements programs and projects for USDA sustainable practices, environmental response and restoration, and biobased product market transformation.

USDA's BioPreferred program increases the development, purchase, and use of biobased products through two initiatives: mandatory biobased product purchasing requirements for federal government agencies (regulatory initiative) and voluntary bioproduct certification and labeling. Both parts of the program work in concert as end-use market tools in USDA's supply chain approach to hastening the global transition to the bioeconomy. An economic impact analysis recently released by the program finds that the biobased products industry contributed a total of $369 billion value to the U.S. economy. Biobased products industries directly employ approximately 1.5 million people, while an additional 2.5 million jobs are supported in other sectors. Environmentally, the increased use of biobased products currently displaces about 300 million gallons of petroleum per year – equivalent to taking 200,000 cars off the road. (All numbers as of 2013.) Numerous efforts are continually underway to increase biobased product procurement federal government-wide, track product use, and evaluate how these products are helping Federal agencies meet their sustainability goals. The more than 2,200 products that have received certification to display the USDA Certified Biobased Product label are creating and increasing consumer and commercial awareness about a material's biobased content as one measure of its environmental footprint.

The two elements of the BioPreferred program work in concert as end-use market tools in USDA's supply chain approach to hastening the global transition to the bioeconomy. Further, DM operations work to make USDA facilities greener and reduce the greenhouse gas footprint of departmental assets.

Economic Research Service

The Economic Research Service (ERS) informs and enhances public and private decision making on economic and policy issues related to agriculture, food, the environment, and rural development.

The ERS strategic plan[8] specifically includes understanding the effects of alternative fuels on agricultural markets, the interactions of commodity, livestock, and food markets with biofuels, bioenergy impacts on rural communities, and the production impacts on natural resources and the environment. Research is focused on these economic analyses and on examining the influence of bioenergy and bioenergy policy on domestic and global agricultural markets, natural resources, the environment, rural communities, and implications for food prices.

To the extent that policies or market developments associated with the bioeconomy affect commodity markets, rural communities, the environment, food prices or other issues within the domain of the ERS mission, ERS disseminates data, information, and special studies that address these interactions. Specifically, as a Principal Federal Statistical Agency, ERS has a long-standing and ongoing role in the collection, analysis, and distribution of market data and forecasts for the commodities that most directly influence, and are influenced by, the bioeconomy. ERS maintains a comprehensive set of economic data on ethanol and biodiesel production, consumption, trade, as well as for feedstock and co-product markets, in addition to other metrics. Economic modeling at ERS evaluates the economic and competitive potential for forest residues and dedicated energy crops (e.g., switchgrass) as feedstocks for electricity generation, under alternative scenarios such as a renewable portfolio standard or policies to reduce carbon dioxide emissions. Longer-term scenarios consider the role of bio-electricity combined with carbon dioxide capture and storage as a negative-emissions technology.

Farm Service Agency

The Farm Service Agency (FSA) equitably serves all farmers, ranchers, and agricultural partners through the delivery of effective, efficient agricultural programs for all Americans.

This Agency manages the Biomass Crop Assistance Program (BCAP)[9] that provides financial assistance to owners and operators of agricultural and private forestland who wish to establish, produce, and deliver biomass feedstocks. As part of the process, the agency does conservation and forest stewardship plans, cost analysis, and has a Programmatic Environmental Impact Statement (PEIS). FSA also has Interagency Agreements with US Forest Service and Bureau of Land Management to support economically viable retrieval of forest residues from the National Forest System (NFS) for delivery to bioenergy at facilities, at a long haul distance from the NFS. In 2015, USDA announced the Biofuels Infrastructure Partnership (BIP), with USDA sourced funding of $100 million in grants, and that to apply states and private partners match the federal funding by a 1:1 ratio. USDA is partnering with 21 states through the BIP to nearly double the number of fueling pumps nationwide that supply renewable fuels to American motorists.

Forest Service

The Forest Service (FS) sustains the health, diversity, and productivity of the nation's forests and grasslands to meet the needs of present and future generations.

---

8 www.ers.usda.gov/about-ers/strateg c-p an.aspx

9 http://www.fsa.usda.gov/FSA/webapp?area_home&subject_ener&top c_and ng

Woody biomass utilization is an important component of the Forest Service Strategic Energy Framework[10], which sets direction and proactive goals for the FS to contribute significantly and sustainably toward resolving U.S. energy resource challenges, by fostering sustainable management and use of forest and grassland resources. The Forest Service is part of the USDA Wood-to-Energy[11, 12, 13] initiative, an interagency effort to expand renewable wood energy development and use. The Forest Service's wood utilization program develops markets to reduce the cost of hazardous fuels treatments, forest management, and restoration activities. The Forest Service provides woody biomass feedstocks from hazardous fuels reduction treatments and restoration activities. FS researchers[14] work to develop sustainable forest biomass management and production systems; conversion technologies for biofuels, biopower, and bioproducts; and information and tools for decision making and policy analysis.

The USDA Forest Service's emphasis on biomass, wood products, and wood energy encourages market development for woody biomass and provides high quality data to inform business, management, and policy decisions. Forest Service activities reduce investor risk, provide for sustainable feedstocks, and develop new products and efficient fuels. These activities significantly contribute to U.S. energy security, environmental quality, and economic opportunity and firmly supporting the bioeconomy.

National Agricultural Statistics Service (NASS)

The National Agricultural Statistics Service (NASS) is responsible for conducting monthly and annual surveys and preparing official USDA data and estimates of production, supply, prices, and other information necessary to maintain orderly agricultural operations. NASS also conducts the census of agriculture which is currently conducted every 5 years.

Credible data is essential for a broad spectrum of uses ranging from simple visual inspection of data or data series to developing analytical frameworks to more in-depth analyses of policies, supply and demand drivers, markets dynamics, and price formation. NASS is the source of much of the data released by USDA and used by industry and analysts. NASS collects and publishes data on feedstocks relevant to bioenergy production and on grain crushings and coproducts.

In 2014 NASS started publishing on a monthly basis a particularly valuable report, Grain Crushings and Co-Products Production which is part of the Current Agricultural Industrial Reports (CAIR) program. Data are collected from all known mills that produce ethanol. In addition, an operational profile was completed for each facility to determine the presence of dry and/or wet alcohol mill and the nameplate production capacity. All operations which will produce alcohol will be selected for the monthly Dry Mill Producers of Ethanol Survey and/or Wet Mill Producers of Ethanol Survey which ask for quantities of grain used as feedstock and co-products produced.

NASS published the first Fats and Oils: Oilseed Crushings, Production, Consumption and Stocks in 2015. The data is a combination of all known crushing, rendering and refining facilities. Prior to the beginning of data collection, operations were contacted to verify that they met the scope of the project. The Fats and Oils: Oilseed Crushings, Production, Consumption and Stocks also publishes data on animal fats.

10 http://www.fs.fed.us/spec a uses/documents/S gned Strateg cEnergy Framework 01 14 11.pdf

11 http://b ogs.usda.gov/2010/10/25/wood-to-energy-efforts-expand ng-restor ng-econom es-and-ecosysems/#more-29613

12 http://heat ngthem dwest.org/wp-content/up oads//Dav d-Atk ns-HTM-20121.pdf

13 http://b ogs.usda.gov/2013/09/13/funds-promote-deve opment-of-rura -wood-to-energy-projects/

14 http://www.fs.fed.us/research/pdf/RD B oenergy Strategy March 2010.pdf

National Institute of Food and Agriculture

The National Institute of Food and Agriculture (NIFA) provides leadership and funding for programs that advance agriculture-related sciences. We invest in and support initiatives that ensure the long-term viability of agriculture.

NIFA seeks to facilitate the development of regional biomass systems for the sustainable production of biofuels and biobased products. Coordinated Agricultural Projects (CAPs)[15] that consist of partnerships between government, industry, non-government organizations, and universities, including 1890 land-grant universities, tribal nation colleges, and Hispanic serving institutions address the integration of the supply chain. The Biomass Research and Development Initiative (BRDI) is a joint effort between the U.S. Department of Agriculture (USDA) and the U.S. Department of Energy (DOE). Priorities are jointly set by USDA and DOE each year. USDA - DOE Plant Feedstock Genomics for Bioenergy[16] - This competitive grant program is committed to fundamental research in biomass genomics, providing the scientific foundation to facilitate use of lignocellulosic materials for bioenergy and biofuels.

NIFA programs support the entire Bioeconomy supply chain at the regional scale. This begins with research, development and deployment for feedstock development and management; logistics including harvesting, storage, and transportation; and conversion processes for biofuels and biomaterials. Supporting this research is a full examination of environmental, economical, and social sustainability of these bioeconomy systems. The mission of each NIFA program connected to the bioeconomy is to reduce risk for stakeholders. NIFA also supports programs that build the future bioeconomy workforce as well as outreach/extension programs that translate research for stakeholder awareness.

Natural Resources Conservation Service

Natural Resources Conservation Service (NRCS) provides America's farmers and ranchers with financial and technical assistance to voluntarily put conservation on the ground, not only helping the environmental but agricultural operations, too.

NRCS provides technical assistance to producers to assess the feasibility of growing and harvesting biomass residues and its impact on environmental sustainability in individual fields and farms. In addition, NRCS Plant Materials Centers evaluate potential feedstocks in cooperation with USDA-ARS and Universities, collect plant data to support conservation planning for biomass crops, provide demonstrations and training on the production of energy crops, and support best management practices for the production of biomass crops.

Office of Energy Policy and New Uses

Office of Energy Policy and New Uses (OEPNU) is part of the Office of the Chief Economist (OCE) and advises the Secretary of Agriculture on the economic implications of policies and programs affecting the U.S. food and fiber system and rural areas. OCE supports USDA policy decision making by analyzing the impact of proposals and coordinating a response among several USDA agencies.

The Office assists the Secretary of Agriculture in developing and coordinating Departmental energy policy, programs, and strategies. Research is currently underway on biodiesel fuels, ethanol fuels, and other sources of biomass energy. Measurement of atmospheric emissions associated with renewable energy also is under study. OEPNU's scope of work also includes renewable electricity and examines the integration of energy with agriculture, the environment, and rural communities.

15 http://www.n fa.usda.gov/newsroom/news/2011news/09281_b ofue _product on.htm

16 http://genom csc ence.energy.gov/research/DOEUSDA/awards.shtm

Rural Development

Rural Development (RD) commits to helping improve the economy and quality of life in rural America. RD programs help rural Americans in many ways.

Rural Development offers technical and financial assistance to help agricultural producers, cooperatives, and other businesses improve the effectiveness of their operations, produce energy as a new cash crop, and process raw agricultural and forestry raw materials into value-added, biobased products. RD programs have funds available to complete energy audits, complete energy efficiency improvements, install renewable energy systems, construct biorefineries, support production of advanced biofuels and much more. USDA Rural Development is at the forefront of renewable energy financing, with options including grants, guaranteed loans and payments. RD programs, authorized through the Agriculture Act of 2014, fall into four main programs under the Energy Title: the Biorefinery, Renewable Chemical, and Biobased Product Manufacturing Assistance Program provides loan guarantees for the development, construction, and retrofitting of commercial-scale biorefineries and biobased product manufacturing facilities; the Repowering Assistance Program provides payments to eligible biorefineries to replace fossil fuels used to produce heat and/or power to operate the biorefineries with renewable biomass; the Advanced Biofuel Payment Program provides payments to producers in order to support and expand production of advanced biofuels refined from sources other than corn kernel starch; and the Rural Energy for America Program provides assistance to agricultural producers and rural small businesses to complete a variety of projects, including renewable energy systems, energy efficiency improvements, renewable energy development assistance, and energy audits.. The mission area has helped contribute to effective public-private partnerships which connect industry with larger Departmental goals contributing to cleaner air and water as well as reducing waste contributing to a strong biobased economy in the U.S.

ENVIRONMENTAL PROTECTION AGENCY

www.epa.gov

Office of Air and Radiation/Office of Transportation and Air Quality

The Office of Air and Radiation/Office of Transportation and Air Quality (OTAQ) protects human health and the environment by reducing air pollution and greenhouse gas emissions from mobile sources and the fuels that power them; advancing clean fuels and technology; and encouraging business practices and travel choices that minimize emissions.

Renewable Fuel Standard Program

The Renewable Fuel Standard (RFS) Program began in 2006 pursuant to the requirements in Clean Air Act (CAA) section 211(o) that were added through the Energy Policy Act of 2005 (EPAct). The statutory requirements for the RFS program were subsequently modified through the Energy Independence and Security Act of 2007 (EISA), resulting in the publication of major revisions in the regulatory requirements on March 26, 2010. The fundamental objective of the RFS provisions under the Clean Air Act is clear: to increase the use of renewable fuels in the U.S. transportation system every year through at least 2022. These fuels must be derived from renewable feedstock, coming from qualifying land and demonstrate they meet certain minimum reductions of Greenhouse Gas Emissions over the fuels they are replacing. There are several categories of fuels specified in the law: Cellulosic biofuels which are required to have 60 percent or greater greenhouse gas (GHG) emissions benefits on a lifecycle basis than the petroleum based fuels they replace; advanced biofuels 50 percent or greater benefit; and conventional biofuels (other than grandfathered facilities) 20 percent or better benefit. Increased use of renewable fuels means less use of fossil fuels, which results in lower GHG emissions over time as advanced biofuel production and use becomes more commonplace. By aiming to diversify the country's fuel supply, Congress also intended to increase the nation's energy security. The law establishes annual volume targets, and requires EPA to translate those volume targets (or alternative volume requirements established by EPA) in accordance with statutory provisions. Since the initial promulgation of the RFS program regulations in 2007, domestic production and use of renewable fuel (also referred to as biofuel) volumes in the U.S. has increased substantially. Renewable fuels include corn starch ethanol, the predominant biofuel use to date, but Congress envisioned the majority of growth over time to come from advanced biofuels as the non-advanced (conventional) volumes remain constant starting in 2015, while the advanced volumes continue to grow. This program is currently the predominant biobased program with direct and significant intersection with the Bioeconomy Vision. The production of these renewable transportation fuels uses feedstocks that do and will likely continue to compete with feedstocks expected to play a role in the broader bioeconomy. As a broader Bioeconomy market develops, there could be market shifts and competition both in terms of feedstock availability, utilization, and diversification. Market economics will drive the Bioeconomy participants to select the most economical / profitable market. Shifts in feedstock competition, cost, availability and finished product values (Fuels, chemicals, energy, food, other) are likely to have impacts, potentially positive and negative, to existing and emerging markets, policies (mandated programs, subsidized programs, etc.). Evaluating the bioeconomy vision in this context is vital, given the additional complexity that is likely under such market circumstances.

Office of Air and Radiation / Office of Atmospheric Policy (OAP)

OAP protects human health and the environment through implementation of cross-cutting atmospheric programs.

Three programs have relevance to the Bioeconomy:

1. **Framework for Assessing Biogenic CO_2 Emissions from Stationary Sources**

    In November 2014, EPA released the second draft of the technical report, *Framework for Assessing Biogenic CO_2 Emissions from Stationary Sources*. EPA engaged in a targeted peer review on the revised Framework with EPA's Science Advisory Board in 2015.

    The intent of the Framework is to evaluate biogenic CO_2 emissions from stationary sources that use biogenic feedstocks, given the unique ability of biogenic material to sequester CO_2 from the atmosphere over relatively short time frames through the process of photosynthesis. It is a methodological framework for assessing the extent to which the production, processing, and use of biogenic material at stationary sources results in a net atmospheric contribution of biogenic CO_2 emissions. Biogenic CO_2 emissions are defined as CO_2 emissions related to the natural carbon cycle, as well as those resulting from the production, harvest, combustion, digestion, fermentation, decomposition, and processing of biologically based materials. The Framework was developed as a policy-neutral framework for assessing biogenic CO_2 emissions from stationary sources - it was not developed as technical guidance in conjunction with any specific policy or program. However, it was designed to be flexible so that decisions on specific technical components can be made to accommodate different policy applications.

2. **AgSTAR**

    The mission of AgSTAR is to promote the use of biogas recovery systems at livestock operations to reduce methane emissions and achieve other environmental benefits through outreach, education, tools and partnerships.

    AgSTAR develops technical resources for farmers, state government representatives and other stakeholders. AgSTAR also participates in outreach events with livestock producers, renewable energy industry leaders, and state and local governments to raise awareness about the benefits of livestock biogas recovery systems and the federal resources available for project planning and implementation. Finally, AgSTAR hosts a partnership program, which brings together representatives of universities, state and local governments, not-for-profits, and other related organizations to share information and encourage implementation of biogas recovery systems.

    EPA is expanding the scope of AgSTAR's national mapping tool (www2.epa.gov/agstar/agstar-national-mapping-tool) to include sources of wasted food and other organic wastes, as well as facilities with capacity to receive these materials, and will be completed by fall 2016. This update will allow organic waste generators to find anaerobic digestion facilities and allow biogas project developers to understand the potential organics available in a given area.

    AgSTAR promotes widespread adoption of commercially available technologies to capture and use methane in the agricultural sector for power, fuel and other products.

AgSTAR has an interagency agreement with USDA to coordinate efforts to promote biogas recovery in the agricultural sector.

USDA, DOE and EPA jointly released the Biogas Opportunities Roadmap in August 2014. The Roadmap builds on progress made to date to identify voluntary actions that can be taken to reduce methane emissions through the use of biogas systems and outlines strategies to overcome barriers limiting further expansion and development of a robust biogas industry in the U.S. EPA participates in the Biogas Opportunities Roadmap Working Group (along with DOE and USDA) to pursue these solutions and enhance Federal communication and collaboration regarding biogas activities.

3. **Landfill Methane Outreach Program (LMOP)**

The EPA's Landfill Methane Outreach Program (LMOP) is a voluntary assistance program that helps to reduce methane emissions from landfills by encouraging the recovery and beneficial use of landfill gas (LFG) as a renewable energy resource. LFG contains methane, a potent greenhouse gas that can be captured and used to fuel power plants, manufacturing facilities, vehicles, homes and more. By finding cost-effective ways to utilize landfill methane as energy, LMOP helps to reduce GHG emissions and prevent air pollution, encourage development of a renewable energy resource, promote local economic development, and reduce dependence on non-renewable fossil fuels.

LMOP forms partnerships with communities, landfill owners, power marketers, states, project developers, tribes, and non-profit organizations to overcome barriers to project development by helping them assess project feasibility, find financing, and market the benefits of project development to the community. LMOP offers a wide array of free technical, promotional, and informational tools as well as services to promote the development of LFG energy projects. Through its partnerships, LMOP creates a vital network of landfills, states, communities, and companies interested in LFG use.

LMOP encourages the development and beneficial use of LFG as a renewable fuel resource.

Office of Air and Radiation/Office of Air Quality Planning and Standards

The Office of Air and Radiation/Office of Air Quality Planning and Standards (OAQPS) preserves and improves the quality of the air that we breathe. To accomplish this, OAQPS compiles and reviews air pollution data; develops regulations to limit and reduce air pollution; assists states and local agencies with monitoring and controlling air pollution; makes information about air pollution available to the public; and reports to Congress the status of air pollution and the progress made in reducing it.

On August 3, 2015, the EPA issued the first-ever national standards to address carbon pollution from existing fossil-fuel fired power plants, also known as the Clean Power Plan. Under the authority of Clean Air Act (CAA) section 111(d), the EPA is establishing a CO_2 emission performance rate for each of two subcategories of fossil fuel-fired Electric Generating Units (EGUs) -- fossil fuel-fired electric steam generating units and stationary combustion turbines – that expresses the "best system of emissions reduction… adequately demonstrated" (BSER) for CO_2 from the power sector. The EPA is also establishing state-specific rate-based and mass-based goals that reflect the subcategory-specific CO_2 emission performance rates and each state's mix of affected EGUs. The guidelines also provide for the development, submittal and implementation of state plans that implement the BSER – again, expressed as CO_2 emission performance rates – either directly by means of source-specific emission standards or other requirements, or through measures that achieve equivalent CO_2 reductions from the same group of EGUs.

The EPA also issues preconstruction (i.e., New Source Review (NSR) permits) and construction (i.e., Title V) permits for sources that use biomass as feedstocks as long as these sources emit or have the potential to emit a non-greenhouse gas pollutant, in addition to greenhouse gases, at or above the applicable permitting thresholds.

OAQPS routinely interacts with various USDA offices, and also, through their Federal Advisory Committee, participates in the Agricultural Air Quality Task Force that provides advice to EPA on air quality issues impacting agriculture.

It is recognized that some biomass-derived fuels can play an important role in CO_2 emissions reduction strategies. It is anticipated that some states may consider biomass-derived fuels used in energy production as a way to mitigate CO_2 emissions, and will include them as part of their state plans to meet the emission guidelines in the Clean Power Plan or as part of their NSR's GHG Best Available Control Technology (BACT) review.

Office of Chemical Safety and Pollution Prevention

Office of Chemical Safety and Pollution Prevention (OCSPP) protects you, your family, and the environment from potential risks from pesticides and toxic chemicals. Through innovative partnerships and collaboration, OCSPP also works to prevent pollution before it begins. This reduces waste, saves energy and natural resources, and leaves our homes, schools and workplaces cleaner and safer.

Toxic Substances Control Act

EPA's Office of Pollution Prevention and Toxics (OPPT) manages programs under the Toxic Substances Control Act (TSCA). Under this law, EPA evaluates new and existing chemicals and their risks, and finds ways to prevent or reduce pollution before it gets into the environment.

The premanufacturing oversight portion of TSCA (Section 5) provides EPA with authority to review all new chemicals and new microorganisms subject to TSCA. These include microorganisms and chemicals used for a variety of applications such as biomass conversion for chemical production, waste and pollution control, and mining and resource extraction. Premanufacturing notification is required prior to commercial production for new substances subject to TSCA (Premanufacturing Notification [PMN] for chemicals, Microbial Commercial Activity Notice [MCAN] for microorganisms) to enable human health and environmental risk assessment [>]. Notification is also required (TSCA Experimental Release Application [TERA]), for all new microorganisms, for research activities that constitute environmental testing.

EPA conditionally exempts certain organisms used in contained systems from MCAN review and conditionally exempts all R&D done inside structures from TERA review, provided specified safety procedures are complied with. Only commercial research is subject to TSCA oversight, but much academic research has commercial intent and is covered.

With these procedures, TSCA oversight provides public confidence that activities in the commercial space that use newer technologies like biotechnology is subject to detailed risk assessment.

Biotechnology throughout the Federal establishment is subject to coordination under the Coordinated Framework for Regulation of Biotechnology, led by the Office of Science and Technology Policy and involving all the relevant regulatory agencies. That framework, last updated in 1992, is undergoing a review started in July 2015.

EPA'S Pollution Prevention (P2) and Green Chemistry Programs

The mission of EPA's Pollution Prevention (P2) Program is to prevent pollution at the source, promote the use of greener substances, and conserve natural resources. The P2 Program's authority comes from the 1990 Pollution Prevention Act, which established a national policy to prevent or reduce pollution at the source, wherever feasible. The Presidential Green Chemistry Challenge Awards Program[17] established in 1996, is one of several initiatives within the EPA's P2 Program. The Presidential Green Chemistry Challenge Awards Program promotes the environmental and economic benefits of developing and using green chemistry. The EPA manages the program in partnership with the American Chemical Society, and in the 20 years the program has been operating has received over 1500 nominations and made 104 awards. Many of the recent awards are for using biobased feedstocks as substitutes for petroleum feedstocks to efficiently produce a wide range of products, from biofuels to biopolymers and specialty chemicals.

On July 13, 2015 EPA and the American Chemical Society held the 20th Annual Presidential Green Chemistry Awards. Among the six winners was an academic researcher who developed a process for using plant-based materials in the production of renewable chemicals and liquid fuels via a waste-free and metal-free catalytic process. One of the industry winners was a company that developed a blue-green algae to produce ethanol and other fuels; the algae uses CO_2 from air or industrial emitters with sunlight and saltwater to create fuel while dramatically reducing the carbon footprint, costs, and water usage, with no reliance on food crops as feedstocks. Another industry winner was a company that developed a safer, plant-based polyurethane for use in a variety of applications, including flooring, furniture, and foam insulation; perhaps most notably, the technology eliminates the use of isocyanate feedstocks. A fourth winner developed a cost-effective two-step process using supercritical water to deconstruct a wide range of renewable plant material and produce separate streams of cellulosic xylose and glucose for use as building block chemicals in a multitude of downstream technologies; after sugar extraction, remaining lignin solids can be burned to supply the bulk of the heat energy required for the process or utilized in higher-value applications like adhesives or thermoplastics.

Other recent awardees include: a vegetable oil (soy) dielectric fluid for electrical transformers that is less flammable, higher performing, and less environmentally risky if spilled, than traditionally used mineral oil; a carbohydrate-based firefighting foam certified for hydrocarbon fire suppression that does not include persistent (and potentially toxic) fluorinated surfactants; and high-throughput methods for using supercritical water (versus acids, enzymes, solvents) to economically break down plant material into low-value sugars that are used as building blocks for renewable chemicals and fuels. These awardees represent just a few of the highly innovative and high impact green chemistry technologies that have received recognition through this competitive awards program.

The EPA also has focused its P2 grant programs to include opportunities for grantees green chemistry innovations into the marketplace, with a focus on improving the environmental and economic performance of small and medium-sized companies. In FY2015, the EPA issued more than 70 grants.

Many of the recent Presidential Green Chemistry Challenge Awards have been for technology innovations that focus on either transforming plant-based materials economically into feedstocks for manufacturing chemicals/feedstocks, or for transforming such feedstocks into valuable biobased products like fuels and/or polymers. These innovations in materials-transformation are key enabling technologies for the Bioeconomy.

---

17 www2.epa.gov/greenchem stry

Many companies that have received the Presidential Green Chemistry Challenge Award have also received grants or loans from other federal programs to promote renewable, biobased economic development. EPA is increasingly looking to work with other federal agencies (e.g., USDA, DOE, NSF) on identifying supply chain opportunities (or barriers) associated with adoption of green chemistry innovations into the marketplace.

Office of Land and Emergency Management

The Office of Land and Emergency Management (OLEM) provides policy, guidance and direction for the Agency's emergency response and waste programs. OLEM develops guidelines for the land disposal of hazardous waste and regulation of underground storage tanks. OLEM provides technical assistance to all levels of government to establish safe practices in waste management. OLEM administers the Brownfields program which supports state and local governments in redeveloping and reusing potentially contaminated sites. OLEM also manages the Superfund program, which responds to abandoned and active hazardous waste sites and accidental chemical releases. Finally, OLEM encourages innovative technologies to address contaminated soil and groundwater.

Office of Underground Storage Tanks

The Office of Underground Storage Tanks (OUST) carries out a Congressional mandate, under Subtitle I of the Solid Waste Disposal Act to develop and implement a regulatory program for underground storage tank (UST) systems (40 CFR Part 280) storing petroleum and certain hazardous substances.

In 1985, EPA created the Office of Underground Storage Tanks to carry out a Congressional mandate to develop and implement a regulatory program for underground storage tank (UST) systems. EPA works with its state, territorial, and tribal partners to prevent and clean up releases from UST systems. Many UST systems are designed for 30 year lifespans or longer, and many existing UST systems have components that were installed before ethanol and biodiesel became commonly blended with gasoline and diesel for use as motor fuel.

Biofuels can be more aggressive towards certain parts of UST systems than their hydrocarbon counterparts, and may affect physical qualities of some components and cause them to swell or shrink or become hard or brittle. This incompatibility could cause a release of fuel to the environment. Biofuel blends also show higher rates of corrosion in UST systems which can prevent UST equipment, including release prevention equipment, from functioning correctly.

The Office of Underground Storage tanks has led several regulatory and research efforts to address biofuel compatibility and corrosion considerations for UST systems. In 2011, the office released guidance for owners and operators to assist them when determining and demonstrating compatibility of their UST systems to ensure their UST system can store biofuels compatibly. New requirements in the updated 2015 federal UST regulation require owners notify implementing agencies of their intent to store biofuels and keep records demonstrating the compatibility of their UST systems. The clarity provided will help prevent biofuels from being stored in systems that are not compatible.

The office has also funded research into: the potential impacts to UST systems of storing E15 blends of ethanol; suitability testing of leak detection equipment in various percentages of ethanol blended fuel; development of protocols for certifying leak detection equipment; corrosion of equipment in USTs storing ethanol blended fuels; and any role of biofuels in internal corrosion of UST systems storing diesel fuel.

The Office of Underground Storage Tanks continues to work to find answers to existing and evolving compatibility and corrosion concerns with biofuels to ensure these fuels can be stored safely in UST systems.

Office of Resource Conservation and Recovery

The Office of Resource Conservation and Recovery (ORCR) is primarily responsible for implementing EPA's resource conservation, recovery and waste management goals under the Resource Conservation and Recovery Act (RCRA.) A principal responsibility is to build a national waste management program, implemented through EPA Regional Offices and State Programs. The Office is responsible for promoting sustainability and safe materials management, and fostering waste reduction and responsible management practices that will conserve natural resources, prevent future problems, and clean up problems from the past. Specific activities include:

Waste Reduction Model (WARM): In March 2015, EPA updated WARM to incorporate food wastes. EPA is currently working on updating the model to add anaerobic digestion as a waste management practice. WARM helps solid waste planners and organizations calculate the GHG benefits of alternative end-of-life waste management decisions. These new updates will allow them to consider the potential impacts of anaerobic digestion of food wastes compared to other optional practices when making decisions for their operations.

Reducing Wasted Food: Reducing the amount of food waste sent to landfills continues to be an opportunity for improvement across the United States. Food is the largest stream of materials in our landfills accounting for 21% of the American waste stream. This large volume of disposed food is a main contributor to the roughly 18% of total U.S. methane emissions that come from landfills, a powerful greenhouse gas. In November 2015, EPA will co-host a Food Recovery Summit in Charleston, SC to gather critical feedback and develop key connections with the private and public players needed to achieve significant wasted food reductions.

The Sustainable Materials Management (SMM) is a systemic approach to using and reusing materials more productively over their entire lifecycles, including production of biofuels, biochemicals, and biopower, when appropriate. It represents a change in how our society thinks about the use of natural resources and environmental protection. By looking at a product's entire lifecycle we can find new opportunities to reduce environmental impacts, conserve resources, and reduce costs.

Office of Water

The Office of Water primarily supports the Clean Water Act, Safe Drinking Water Act, and portions of other Federal statutes. Many long-standing regulations and programs, and new initiatives are supportive of the bioeconomy, as doing so positively impacts our nation's waters.

Within the Office of Water's Office of Wastewater Management (OWM), primary regulatory duties include oversight of the regulation of industry and municipal wastewater discharges, in which most states are delegated to carry out the day-to-day aspects. Other activities within OWM include management of the Clean Water State Revolving Fund (CWSRF), which may provide financial assistance to support the bioeconomy through funding projects that produce biofuels and biopower from treatment of municipal wastewater. Many of these projects would be eligible for funding under the CWSRF's Green Project Reserve (GPR), a provision which requires, to the extent sufficient projects are available, that a portion of each state's capitalization grant be provided for green infrastructure, energy and water efficiency, and other environmentally innovative projects. Drinking Water State Revolving Funds (DWSRFs), managed by OW's Office of Groundwater and Drinking Water, are also eligible to be used in non-traditional ways that may support the bioeconomy. If funded by Congress, the new Water Infrastructure Finance and Innovation Act (WIFIA) program is another useful tool that will be able to provide low-cost financing for any project already eligible under the CWSRF or DWSRF programs. Through the use of these innovative financing tools, EPA hopes to find more integrated, lower long-term cost solutions to address the estimated $600 billion in infrastructure needs over the next 20 years.

Within OWM's Rural Branch, several voluntary collaborative initiatives have been launched that focus on the animal agriculture industry, such as energy and nutrient recovery from animal manures, for example. The Rural Branch is partnering with dairy and swine producers, USDA, industry trade associations, and academic experts to launch an innovation challenge to encourage the development and adoption of affordable and effective technologies that can extract and transform nutrients in dairy and swine manure (raw and/or digested) and sequester them into reusable and/or sellable products. The challenge is anticipated to launch in fall 2015. Following a call for submissions of technology concepts, EPA and collaborators will implement multiple phases for innovators to bring their concepts through technology designs, then prototypes, and ultimately to pilot-scale systems to be demonstrated at U.S. dairy and swine operations.

Annually, more than 2 billion livestock animals at animal feeding operations (AFOs) and concentrated animal feeding operations (CAFOs) generate over a billion tons of manure[18]. Manure can be used as a feedstock in renewable energy generation systems, such as anaerobic digesters and gasifiers, to generate biogas and synthetic natural gas (syngas). Manure can also be transformed into a range of value-added co-products, such as composts, potting soils, animal bedding, paper products, planting pots, particle boards, and concentrated nutrient products.

A few U.S. livestock operations (10, most of which are poultry operations) are experimenting with gasification and combustion systems. Such combustion systems that are able to concentrate the majority of nutrients in manure in a mineral-rich ash product, which may be used as a fertilizer and soil amendment.

In the past decade, innovators have developed and demonstrated technologies that can recover a significant amount of N and P from manure, as well as from wastewater treatment plants (now being referred to as water resource recovery facilities [WRRFs]), and yield concentrated, dry, and potentially valuable nutrient products, such as struvite and ammonium sulfate fertilizer.

OWM regularly coordinates with EPA AgSTAR, the Office of Research and Development, as well as USDA's Agricultural Research Service (ARS), Natural Resource Conservation Service (NRCS), and Rural Development (RD) on development of innovative manure nutrient recovery technologies. OWM also coordinates with EPA's AgSTAR Program and the Office of Land and Emergency Management, USDA's Rural Development, and DOE's Bioenergy Technologies Office on development of the Biogas Roadmap, which supports the President's Climate Action Plan.

Within OW's Office of Wetlands Oceans and Watersheds (OWOW), the Nonpoint Source Branch provides grants (known as 319 Grants) that are targeted towards non-point sources (i.e., other than regulated entities under the Clean Water Act). These grants may also be used in ways that support the bioeconomy. OWOW is responsible for projection of our nation's wetlands.

OW as an overall office collaborates with many other federal agencies in issues that touch upon the bioeconomy, such as Department of Energy on energy and nutrient recovery from municipal waste water, and agencies active in climate adaptation.

---

18 USDA. 2014. 2012 Census of Agr cu ture, USDA. 2005. FY-2005 Annua Report Manure and Byproduct *Utilization National Program 206.*

Office of Research and Development

The Office of Research and Development (ORD) conducts leading-edge research and fosters the sound use of science and technology to fulfill EPA's mission to protect human health and safeguard the natural environment. ORD develops flexible tools that inform decision-making based on a holistic assessment of impacts to environmental, social, and economic dimensions enabling insights into the trade-offs associated with alternative decision options.

Net Zero

As ORD's Net Zero Initiative moves forward, a research area of particular interest is the nexus between Net Zero/Net Positive energy and Net Zero waste. Net Zero scientists are looking into the development and implementation of innovative technologies that recover waste streams to produce energy, like co-digestion. In the process of co-digestion, food waste, oils and grease are added to under-capacity anaerobic digesters in wastewater treatment plants to produce energy in the form of biogas.

This project links to the Bioeconomy by providing a sustainable method to utilize biomass resources for energy production, and will support ORD's mission to transition the term "waste" to "resource".

The Net Zero team is partnering with the U.S. Army and the U.S. Army Corp of Engineers to investigate the feasibility of co-digestion at Fort Huachuca, AZ.

National Research Area - Sustainable and Healthy Communities

The EnviroAtlas is a multi-scaled spatial data resource and analysis tool that covers the conterminous US. Focused on mapping the supply and demand of ecosystem services (with over 300 national-scale GIS coverages), the EnviroAtlas also contains coverages depicting drivers of change (e.g. population, land use, resource exploitation, pollution, projected scenarios of climate change), demographics, economic data (e.g. employment dependent on ecosystem services), as well as features of the built environment (e.g. transportation). It is scalable from national to regional to local and includes analysis tools that allow evaluation of "what-if" scenarios. The EnviroAtlas is a useful tool for identifying sensitive resources (habitat for vulnerable species, limited water supplies) as well as opportunities for development. Future scenarios for land use and climate change can be used to explore possibilities of co-benefits or unintended consequences.

The EnviroAtlas is a multi-agency/multi-institution effort including collaboration with USFS, USGS, NRCS, NatureServe, and a number of academic institutions. It is a foundational component of the Administration's EcoInforma effort.

National Research Area - Air, Climate, and Energy

ORD conducts leading-edge research and fosters the sound use of science and technology to fulfill EPA's mission to protect human health and safeguard the natural environment. The Air, Climate, and Energy (ACE) Research Program conducts research to evaluate the potential environmental impacts of increased biofuel production and use. Research is examining the effects of changing biofuel content on vehicle emissions and toxicity, the potential change in nutrients associated with increased biofuel feedstock production, and changes in emissions due to co-firing of coal and biomass for power generation. Additional work addresses potential changes in environmental impacts across the full life cycle of production and use for biomass- and fossil-fueled energy.

The Program's research is designed to inform decision makers about the potential positive and negative impacts of large-scale adoption of biomass-based energy.

Work is coordinated with USDA to understand how increased biomass production may affect feedstock prices and with DOE to understand advances in biofuel conversion technologies.

National Research Area - Safe and Sustainable Water Resources

ORD through the Safe and Sustainable Water Resources (SSWR) research program includes projects that focus on the costs of nutrient release to the environment and systems research on best management practices for nutrient sources. As the bioeconomy relies on agricultural biomass as feedstock, SSWR nutrient research informs practitioners in multiple sectors (e.g., irrigation, cropland management, CAFO, wastewater, and transportation) about the potential impacts or benefits of agricultural practices in protecting water quality and watersheds. This work represents collaborative research with USGS and USDA.

The SSWR program also focuses on water reuse and resource recovery. SSWR conducts research on treatment technologies for fit-for-purpose water with an emphasis on low cost/low energy treatment systems. Resource recovery from wastewater and water reuse treatment aims to maximize energy recovery from biosolids through the generation and capture of methane.

EPA Regional Offices

Region 4

Regionally Applied Research Effort (RARE): Through a RARE project Region 4 provided funds to Mississippi State University in a research project that demonstrated the use of algae as a feedstock for biofuels. Region 4 is currently coordinating with EPA's ORD on a RARE project to investigate the use of common agricultural wastes in Region 4, such as corn stover and forestry wastes, in a process that will result in torrefied pellets that can be used as a bio-coal supplement to fossil coal as well as a bio-char that can be used as a soil amendment. The study will determine if the bio-coal pellets will have an equivalent heating content as fossil coal while reducing emission levels of controlled pollutants such as NOx.

Southeast Regional Partnership for Planning and Sustainability (SERPPAS): Region 4 is a member of SERPPAS which is a six-state partnership of state and federal agencies that promotes collaboration in making resource-use decisions supporting conservation of natural resources, working lands, and national defense. Specifically regarding biofuels, the Region coordinates the partnership's Biomass Workgroup. This Workgroup supports the sustainable use of forest resources to produce energy in order to support local economies, create clean energy jobs, reduce energy related greenhouse gas emissions, and enhance energy independence and security. The Workgroup's goals include advancing the use of forest biomass to energy initiative, bringing Federal and state agencies together to identify, address and communicate current and emerging issues in the use of biomass for energy, sharing information on successes and lessons learned regarding biomass energy, and increasing the range of technologies available for biomass projects in the region.

Region 4's Southeast Diesel Collaborative (SEDC) was established to reduce emissions from the existing diesel fleet, such as through the use of biodiesel. The SEDC has been very proactive in promoting the production and use of biodiesel and has coordinated with the Department of Energy's National Renewable Energy Laboratory as well as that Departments' Clean Cities programs. The SEDC has hosted two biodiesel workshops that discussed the intricacies of the production of biodiesel as well the implementation of biodiesel strategies for fleet managers. Biodiesel producers such as St. Johns County, Florida and Hoover, Alabama have been frequent speakers at SEDC conferences and workshops.

Region 4 has promoted the use of biomethane as a transportation fuel. The Region has coordinated tours of the Seminole Landfill biofuels facility with Dekalb, County, Georgia. State and local air directors as well as other EPA staff have toured this facility which collects landfill gases and cleans them to pipeline quality so they may be used in the County's waste haulers as well as being sold to the public as a transportation fuel. Additionally, representatives from the facility have been frequent speakers at Regional conferences and meetings.

Region 4 representatives served on the State of Georgia's "One Stop Shop" for biofuels. This was an effort by the State to provide a forum for biofuel and energy companies considering locating in Georgia to simultaneously meet with all the representatives from state and federal agencies having jurisdiction on biofuel and energy production facilities. The goal is to expedite an understanding of necessary legal requirements.

Region 4 has worked to identify and promote the availability of alternate fuels along critical freight corridors such as the Region 4 portion of the I-75 (the Green Corridor project). Additionally, the Region has coordinated with the State of Tennessee in implementing its Green Corridors project for alternative fuels throughout interstates in that State. Region 4 has also worked with Colonial Pipeline in their efforts to ship biodiesel blends on its pipeline to central and southern Georgia.

The Region has been very active in promoting biodiesel. The Region worked with the National Park Service to promote and provide technical advice for the use of biodiesel in Park Service vehicles. The Region also coordinated with the Cherokee Indian Nation on the use of biodiesel and provided funds for the implementation of an overall biodiesel program that included two transit buses and six school buses. The Region has also provided grants to North and South Carolina for biodiesel fleets, including freight locomotives in and around the Port of Charleston. Additionally the Region has coordinated with non-profit organizations such as the Southern Alliance for Clean Energy (SACE) to promote the use of biofuels and the establishment of biofuel refueling networks.

Region 9

EPA Region 9 provided funds to demonstrate an ultra-low nitrogen oxides (NOx) biogas powered engine installed at a dairy farm in the San Joaquin Valley, California. More information can be found at the Air District website.[19] EPA Region 9 funded refuse trucks fueled by renewable natural gas at a food waste facility in Sacramento, California. Through anaerobic digestion, food waste is converted into compressed natural gas to fuel refuse trucks and provide electricity and heat used at the facility. The remaining processed material is turned into fertilizer and soil amendments.[20]

EPA ORD and Region 9 are conducting research to compare different biogas management technologies by examining air quality, greenhouse gas and economics/operations. This analysis may enable governments, regulators and project developers to identify geographically-appropriate and cost-effective biogas management options. DOE and USDA's NRCS will continue to be involved with this research.

---

[19] http://va_eya r.org/grants/documents/techno ogyadvancement/C-4236_EF&EE_F na Report.pdf .

[20] http://westcoastco aborat ve.org/fi es/grants/smaqmd-dera-fy13-refuse-truck-ag-tractor-rep acements-wcc-factsheet.pdf

EPA Region 9 is publishing food recovery guides for six metropolitan areas in the Pacific Southwest, which will include listings for anaerobic digestion and fats, oils, and grease recycling opportunities. The guides are intended for food waste-generating businesses to facilitate greater food recovery. The guides are expected to be published November, 2015.

In March 2015, EPA published a report titled "Food Waste to Energy: How Six Water Resource Recovery Facilities are Boosting Biogas Production and the Bottom Line."[21] This report presents the co-digestion practices, performance, and experiences of six such water resource recovery facilities. The report describes the types of food waste co-digested and the strategies—specifically, the tools, timing, and partnerships—employed to manage the material. Additionally, the report describes how the facilities manage wastewater solids, providing information about power production, biosolids use, and program costs.

Starting in 2012, EPA Region 9 convened a state/federal/local working group to facilitate the proliferation of dairy digesters in California, which is the nation's largest dairy producer. EPA co-chaired the working group along with NRCS and California Department of Food and Agriculture, and the group included all relevant state and local agencies. The group accomplished several things including: 1) developing a coordinated permitting system of dairy digesters across the phalanx of agencies in California, 2) getting digesters included in the renewable energy mix required by California, and 3) developing criteria for the types of digesters that are ideal to be built in California. In 2014, $12 million dollars from the California Department of Food and Agriculture was awarded to dairy digester projects in California as a result of this working group.

---

21 epa.gov/reg on9/organ cs/ad/epa-600-R-14-240-food-waste-to-energy.pdf.

DEPARTMENT OF INTERIOR

www.doi.gov

Bureau of Land Management

The Bureau of Land Management (BLM) manages and conserves the public lands for the use and enjoyment of present and future generations under our mandate of multiple-use and sustained yield.

Nearly one fourth of the lands - 58 million acres - managed by the BLM are forests or woodlands. An estimated 14 million acres have been affected by the suppression of natural fire and are in need of active management to restore resilient species composition and stand structure. Forest and woodland management produces traditional wood products as well as woody biomass which primarily results from restoration residues and smaller diameter material from forestry, fuels and rangeland treatments.

Biomass from BLM forest and woodland projects has become part of the feedstock that energy companies are relying on to meet various State and Federal renewable energy portfolio standards.[22]

From 2010 to 2014, BLM sold an average of 177,000 tons of biomass for energy through contracts and permits.[23] Additionally the BLM entered into a Memorandum of Agreement in 2014 with the USDA Farm Services Agency to implement the Biomass Crop Assistance Program (BCAP) that allows collection, harvest, storage, and transportation assistance for biomass delivered to certified bioenergy facilities from BLM lands. BCAP projects on BLM land meet environmental standards equivalent to forest stewardship plans and the biomass is limited to material generated as a byproduct of wildfire hazard fuels treatments or treatments to reduce or contain disease or insect infestations in accordance with the Agricultural Act of 2014.[24]

22 https://www.do .gov/s tes/do .opengov. bmc oud.com/fi es/up oads/FY2016_BLM_Greenbook.pdf

23 https://www.do .gov/s tes/do .gov/fi es/m grated/pmb/ppp/up oad/DO -APPR-02022015-v2-F na .pdf

24 http://www.fsa.usda.gov/ nternet/FSA_F_e/ccc_2015_0001_0001.pdf

NATIONAL SCIENCE FOUNDATION

www.nsf.gov

The Division of Chemical, Bioengineering, Environmental, and Transport Systems (CBET) supports innovative research and education in the fields of chemical engineering, biotechnology, bioengineering, and environmental engineering, and in areas that involve the transformation and/or transport of matter and energy by chemical, thermal, or mechanical means.

Standing programs within the CBET Division of the ENG Directorate, including the Energy for Sustainability, Environmental Sustainability, Biochemical Engineering, Catalysis & Biocatalysis, and Process & Reaction Engineering programs, provide funding opportunities for supporting fundamental scientific research needed to develop future processes for advancing the bioeconomy. Awards typically range from $300-400k for 3 years. There are no solicitations or proposal calls specifically for the bioecomony within CBET. However, many of unsolicited proposals, submitted primarily from the academic research community, are closely related to topics relevant to the bioecomony, as described below. Proposals on bioeconomy topics that are the most strongly recommended for funding through the peer review processes are considered for funding. CAREER awards offered through each program are designed to support early career, tenure-track faculty in these areas.

The standing programs identified above support research relevant to advancing the bioecomony in the conversion and sustainability areas through development and scientific understanding of 1) new chemical, catalytic, and biological processes for the conversion of renewable carbon resources (plant biomass, algae, CO_2) to biofuels, bioproducts, biochemicals, as well as their downstream separation; 3) new catalysts and synthetic biology approaches for supporting these processes; 3) biopower production though energy-positive waste water treatment, 4) new tools for life cycle, environmental, and sustainability analyses of biofuel and biorefinery processes.

Engineering Directorate, Emerging Frontiers in Research and Innovation

The Emerging Frontiers in Research and Innovation (EFRI) provides critical, strategic support of fundamental discovery at the frontiers of engineering research and education.

Each year, EFRI develops and issues two topics for through a formal solicitation process. Up to $16 million is available to support each topic on up to 8 awards of approximately $2 million each. Proposals are recommended for funding through a highly-competitive peer review process.

Recent EFRI solicitations relevant to the bioeconomy in the conversion and sustainability areas were HYBI (NSF 08-599, Hydrocarbons from Biomass, FY09), and PSBR (NSF 12-583, Photosynthetic Biorefineries, FY13). The HYBI program supported research on new and potentially transformative approaches for chemical and catalytic conversion of biomass constituents and algae to energy-dense molecules and drop-in hydrocarbon biofuels.

The PSBR program supported fundamental research to advance development and scientific understanding of biofuel and biorefinery processes by linking fundamental biology of phototrophic organisms, bioprocess engineering, and life cycle analysis through a multiscale analysis approach.

Biological Sciences Directorate

The Biological Sciences Directorate (BIO) Directorate programs support research advances at the frontiers of biological knowledge, increases understanding of complex systems, and provides a theoretical basis for original research in many other scientific disciplines.

The Plant Genome Research Project (PGRP) within the BIO Directorate supports academic research for the development and fundamental scientific understanding of functional genomics tools and sequence resources for use in the study of key crop plants and their models. From FY12 to FY14, the program supported approximately $120 million in research.

The Plant Genome Research Project supports the bioeconomy in the feedstocks area by advancing the development and genetic tools for key crop plants and their models.

Science, Engineering and Education for Sustainability

Science, Engineering and Education for Sustainability (SEES) is a cross-cutting initiative within NSF to advance science, engineering, and education to inform the societal actions needed for environmental and economic sustainability and sustainable human well-being. The SEES portfolio activities highlights NSF's unique role in helping society address the challenge(s) of achieving sustainability. Recent special solicitations and funding opportunities to support the bioeconomy in the context of SEES included:

The Sustainable Energy Pathways (SEP) solicitation (NSF 11-590, FY12) supported fundamental research on scalable approaches for sustainable energy conversion to useful forms, as well as its storage, transmission, distribution, and use, through a systems approach to designed to seek comprehensive understanding of the scientific, technical, environmental, economic, and societal issues.

The Sustainable Chemistry, Engineering, and Materials (SusChEM) Funding Opportunity (NSF 15-085) addresses the interrelated challenges of sustainable supply, engineering, production, and use of chemicals and materials.

The SEP program supported 7 projects to advance the bioeconomy through fundamental research on biofuel production platforms in the context of their integrated scientific, technical, environmental, economic, and societal issues. The SusChEM initiative advanced the bioeconomy by supporting a variety of projects related to the sustainable production of biofuels, bioproducts, and biochemicals through sustainable carbon resources.

DEPARTMENT OF TRANSPORTATION

www.dot.gov

Federal Aviation Administration

The Federal Aviation Administration (FAA) strives to provide the safest, most efficient aerospace system in the world. FAA has taken a comprehensive approach to overcome barriers to the development and deployment of sustainable alternative (e.g., renewable) jet fuels that are drop-in replacements to fuels derived from petroleum, as part of an effort to achieve carbon-neutral growth at 2005 emissions levels starting in 2020. These efforts include R&D via the Continuous Lower Energy Emissions and Noise (CLEEN) and Aviation Sustainability Center (ASCENT) University Center of Excellence program programs to evaluate alternative fuels for safety, technical, and environmental performance; sponsorship with industry of the Commercial Aviation Alternative Fuels Initiative (CAAFI), a broad stakeholder coalition; new fuel certification through ASTM International; a partnership with USDA, DOE, and the aviation industry to enable fuel supply development via the Farm to Fly 2.0 Initiative; and the joint development of a federal alternative jet fuel R&D strategy in partnership with numerous federal agencies. FAA also has bilateral agreements to leverage other countries' efforts to develop, assess, and deploy alternative jet fuels, and participates in the United Nation's International Civil Aviation Organization (ICAO) activities on alternative jet fuels.

FAA's goal of goal of having 1 billion gallons of alternative jet fuel in use by 2018 has spurred rigorous research and rapid innovation in advanced feedstock and renewable aviation fuel pathway development. FAA-supported testing and demonstrations of many biobased jet fuels have shown a sharp reduction of air quality emissions with no adverse impact on aircraft engine operations. Broad deployment of renewable jet fuels may substantially reduce lifecycle aircraft exhaust emissions of aviation operations, and enhance public awareness of advanced biofuels.

National Highway Transportation Safety Administration

National Highway Transportation Safety Administration (NHTSA) works daily to help prevent crashes and their attendant costs, both human and financial.

Vehicle Corporate Fuel Economy (CAFE) standards are regulated by NHTSA; NHTSA sets and enforces the CAFE standards, while the Environmental Protection Agency (EPA) calculates average fuel economy levels for manufacturers, and also sets related GHG standards. NHTSA has proposed to require badges, labels and owner's manual information for new passenger cars, low-speed vehicles (LSVs) and light-duty trucks in order to increase consumer awareness regarding the use and benefits of alternative fuels, including biofuels.

E85 capable flex-fuel vehicles (FFVs) have long-qualified for a special calculation of their fuel economy performance under NHTSA administered CAFE regulations. These special calculations provide vehicle manufacturers with powerful credit incentives to develop and produce FFVs. The existence of the current national FFV fleet is largely attributable to this regulatory incentive (effective in its current form through model year 2016). Consumer education on biofuel end-use would be advanced by NHTSA's proposed rule to require vehicle OEMs to prominently exterior-label vehicles capable of running on biofuels, describe the capabilities and benefits of using alternative biofuels to owners' manuals provided for alternative fuel vehicles, and clearly label the fuel filler compartment of vehicles capable of running on biofuels.

Federal Highway Administration

The Federal Highway Administration (FHWA) strives to improve mobility on our Nation's highways through national leadership, innovation, and program delivery.

The Congestion Mitigation and Air Quality (CMAQ) Improvement Program provides funding to state departments of transportation (DOTs), municipal planning organizations (MPOs), and transit agencies for projects and programs in air quality non-attainment and maintenance areas that reduce transportation-related emissions, including alternative fuels deployment. The FHWA Office of Real Estate Services has been cooperating with states to examine the both the potential and implications for bioenergy production within highway right-of-ways (ROWs).

Past and current CMAQ projects have provided for the purchase and development of alternative fuel refueling infrastructure, and conversion of public fleet vehicles to operate on alternative fuels, including biofuels. Generating biofuel feedstocks on highway ROWs would add highly-visible bioenergy production capacity, advance federal and state sustainability goals, support a local green job market, and help reduce highway maintenance costs while potentially generating additional revenue for transportation agencies.

Pipeline and Hazardous Materials Safety Administration

Pipeline and Hazardous Materials Safety Administration (PHMSA) works to protect people and the environment from the risks of hazardous materials transportation.

PHMSA works with other Federal agencies, industry, standards organizations, and emergency responders, to ensure adequate design and operating standards for biofuel pipelines. PHMSA has examined the potential for ethanol-induced stress corrosion cracking in existing pipeline infrastructure used to transport ethanol and various ethanol-blended fuels. PHMSA also supports emergency responder education and training on optimal emergency response to spill incidents involving ethanol and other biofuels in transportation. In partnership with the Federal Railroad Administration (FRA), PHMSA has undertaken a comprehensive, system-wide approach to prevent and mitigate rail accidents involving flammable liquids, including biofuels. The joint-approach focuses on improvements to rail operations, ensuring proper classification of hazardous materials, and improving tank car survivability.

Safe and efficient pipeline and rail transport will be critical to bioeconomy expansion. Continued federal efforts in research and development, resolving technical issues, and setting/clarifying safety standards are key to reducing transport accidents, mitigating the consequences of an incident, and ensuring proper emergency response.

Maritime Administration

The Maritime Administration (MARAD) promotes the use of waterborne transportation and its seamless integration with other segments of the transportation system, and the viability of the U.S. merchant marine.

MARAD has been evaluating the use of renewable diesel fuel blends in commercial vessels. This has included recent at-sea testing and reporting on engine performance, fuel economy, air emissions, engine vibration, underwater radiated noise, and effect on the engine itself. MARAD has also conducted tests to determine the effects of long-term storage on renewable diesel blend quality.

Successful operational and storage testing is critical for advancing the deployment of biofuels in the waterborne transportation sector and the national merchant marine fleet.

Federal Transit Administration

Federal Transit Administration (FTA) strives to ensure personal mobility and America's economic and community vitality by supporting high quality public transportation through leadership, technical assistance and financial resources.

FTA has supported the expanded use of biofuels (i.e., biodiesel) within the nation's transit bus fleet through a large number of field studies, fleet demonstrations, and funded deployments.

Widespread evaluation of biodiesel blend use among U.S. transit agency fleets generates key information regarding the impact of biofuels on transit bus engines, fuel systems, operations, and maintenance, and has helped establish industry best practices. Successful, highly-visible biodiesel transit bus deployments have helped raise the public's awareness of biofuels and their benefits.

Federal Railroad Administration

Federal Railroad Administration (FRA) works to enable the safe, reliable, and efficient movement of people and goods for a strong America, now and in the future.

FRA has been evaluating the use of biodiesel and renewable diesel fuel blends, as well as biobased lubricants in freight and passenger locomotives. FRA-sponsored feasibility studies have assessed biofuel use impacts on locomotive performance, fuel efficiency, exhaust emissions, and engine parts. In partnership with the Pipeline and Hazardous Materials Safety Administration (PHMSA), FRA has undertaken a comprehensive, system-wide approach to prevent and mitigate rail accidents involving flammable liquids, including biofuels. The approach focuses on improvements to rail operations, ensuring proper classification of hazardous materials, and improving tank car survivability.

Successful feasibility testing of biofuels in locomotives is critical for advancing the deployment of biofuels in the rail transportation sector and national rail passenger and freight fleet. Expanded rail tank car transport of biofuels is also critical to the growth of the bioeconomy, underscoring the need for rail transport safety improvements to prevent accidents, mitigate consequences in the event of an accident, and support emergency response.

Office of the Assistant Secretary for Research & Technology

Office of the Assistant Secretary for Research & Technology (OST-R) is committed to transform transportation by expanding the base of knowledge to make America's transportation system safer, more competitive and sustainable.

OST-R sponsors research focusing on the safe use of new fuels and energy technologies within the transportation system (including biofuels), and the safe, efficient distribution of biofuels and other energy resources across the transportation network. OST-R coordinates research across DOT modal administration, including efforts related to biofuels and bioenergy. OST-R supported research includes the Sun Grant Initiative Regional Competitive Grants Program, which funds research and demon-stration activities related to biomass feedstock, biofuels, and biobased products, and analyses to determine the energy and environmental impacts of bioenergy. The Volpe National Transportation Systems Center, a part of OST-R, conducts model and tool development to analyze transportation needs and constraints associated with biofuel biomass collection, processing, and distribution.

OST-Rs efforts to encourage bioenergy research collaboration between government agencies and regional land grant colleges and universities helps foster innovations, and bioeconomy workforce development. Modeling work by the Volpe Center helps agencies and bioeconomy industry stakeholders better understand how multimodal transportation capacity and infrastructure constraints impact biomass feedstock and biofuel movements (and vice-versa), transport-related environmental impacts, and economic costs. Future policies, infrastructure investments, and other decisions supporting the bioeconomy may hinge upon robust scenario analysis models and tools.

Appendix II: Interagency Activities Supporting the Bioeconomy

As discussed in Appendix I, existing foundational pillars of the bioeconomy are in place to support further expansion Each agency focuses on their own priorities and mission areas, but to fully develop a strong national bioeconomy, synchronization across the government is critical Such examples of these interagency activities include EPA s Renewable Fuel Standard, the Farm to Fly 2 0 Initiative, the Defense Production Act Initiative, the Coordinated Agriculture Projects, the Farm to Fleet Initiative, the Rural Energy for America Program, the Biogas Roadmap, and the Biomass Crop Assistance Program

Feedstock Partnerships across the Government

The U.S. Department of Agriculture (USDA) and the Department of Energy (DOE) have compatible goals in the developing and promoting the use of biomass for biofuels, biopower, and biobased products.[25, 26, 27, 28, 29] Collectively the two Departments have a well-established portfolio covering agronomics, silviculture, genetics, engineering, economics, modeling, and many other scientific disciplines and resource areas. Several USDA agencies have research, technology, deployment or assistance programs associated with various aspects of the production and use of biomass from America's agricultural and forestry lands. DOE has feedstocks development and logistics RDD&D activities distributed across a program, an office, and an agency. USDA has feedstocks plant materials and genetic development, sustainable management and production systems development, logistics, improving interfaces in the supply chain, and integration with conversion activities. Much of the federal government's expertise in these areas is within the Departments and RDD&D activities are completed with National Laboratories, university and industrial partners.

A long standing collaboration between USDA and DOE is the Plant Feedstock Genomics for Bioenergy Program. Initiated in 2006, the program funds fundamental science projects that accelerate plant breeding programs and improve biomass feedstocks by characterizing the genes, proteins, and molecular interactions that influence biomass production.   In addition, the two agencies jointly fund a competitive program in support of the development of a biomass-based industry through the Biomass Research and Development Initiative. The technical areas under consideration include feedstocks development, biofuels and biobased products development, and biofuels development analysis. Projects span research through early stage demonstration.

25 http://energy.gov/quadrenn a -techno ogy-rev ew

26 http://www1.eere.energy.gov/b omass/pdfs/mypp apr 2011.pdf

27 http://www1.eere.energy.gov/b omass/pdfs/b omass two pager.pdf

28 http://www.wh tehouse.gov/s tes/defau t/fi es/rss v ewer/grow ng amer cas fue s.PDF

29 http://sc ence.energy.gov/~/med a/ber/pdf/B ofue s strateg c p an.pdf

The Federal Working Group, "Woody Biomass Utilization Group" led by USDA, DOE and DOI with members also including EPA and DOD promotes and supports the utilization of woody biomass and woody biomass products and residues from forest and woodland health, management and restoration treatments wherever environmentally, economically and legally appropriate. The group has annual plans of work and many accomplishments including a resource center, special projects, and partnering with stakeholders.

Waste Resource Development

USDA, DOE and EPA jointly released the Biogas Opportunities Roadmap in August 2014. The Roadmap builds on progress made to date to identify voluntary actions that can be taken to reduce methane emissions through the use of biogas systems and outlines strategies to overcome barriers limiting further expansion and development of a robust biogas industry in the United States. EPA participates in the Biogas Opportunities Roadmap Working Group (along with DOE and USDA) to pursue these solutions and enhance Federal communication and collaboration regarding biogas activities. A Progress Report is expected to be published in fall, 2015.[30]

Additionally, EPA's AgSTAR program partners with USDA to promote the use of biogas recovery systems at livestock operations to reduce methane emissions and achieve other environmental benefits through outreach, education, tools and partnerships. AgSTAR develops technical resources for farmers, state government representatives and other stakeholders. AgSTAR also participates in outreach events with livestock producers, renewable energy industry leaders, and state and local governments to raise awareness about the benefits of livestock biogas recovery systems and the federal resources available for project planning and implementation. AgSTAR hosts a partnership program, which brings together representatives of universities, state and local governments, not-for-profits, and other related organizations to share information and encourage implementation of biogas recovery systems.[31]

NSF, EPA and DOE jointly hosted the *Energy-Positive Water Resource Recovery (EPWRR) Workshop* in April of 2015 to envision a transition from the wastewater treatment facilities of today to a new generation of Water Resource Recovery Facilities (WRRFs), nationwide, and identify specific opportunities to stimulate and support this transition. Future WRRFs could effectively manage more diverse waste streams, generate biogas for a number of biofuel platforms, produce water and fertilizer, and help communities recover other valuable resources. Participants of this collaborative interagency workshop provided information to federal stakeholders about ongoing industry efforts and how federal activities could best amplify and help realize the industry vision for the WRRF of the Future that would continue to assign top priority to wastewater treatment for the protection of public health and the environment, but also expand its slate of services and products in support of healthy, economically vibrant communities.

Looking Beyond Ground Transportation

In 2014, the U.S. Departments of Energy, Navy, and Agriculture announced three awards to construct commercial scale integrated biorefineries capable of producing renewable jet and diesel that is compatible with military fuel specifications. The partnership between the federal government and private industry to make cost-competitive drop-in biofuels a reality while strengthening the nation's capability to lead the global clean energy market and diversifying potential fuel sources for the Department of Defense. The feedstock used to power these facilities comes from municipal solid waste, woody biomass,

30 B ogas Opportun t es Roadmap: http://www3.epa.gov/c matechange/Down oads/B ogas-Roadmap.pdf
31 For more nformat on on AgSTAR p ease v s t: http://www2.epa.gov/agstar

and waste oils and fats, and will be converted into more than 100 million gallons of renewable jet and diesel fuel annually. This means many additional professional jobs for engineers, plant operators, construction workers, and feedstock suppliers. The renewable jet fuel and diesel that will be produced from these facilities has been extensively tested and qualified by the Department of Defense and proven operate seamless with existing infrastructure, engines, and military operations.[32]

To date fuel blends consisting of up to 50% synthetic hydrocarbons made via the Fischer-Trospch and Hydroprocessed Esters and Fatty Acids pathways are acceptable in military fuel specifications. Additional pathways are being tested. Following adoption into fuel specifications, the Department of Defense has now moved to accept drop-in biofuels as part of its worldwide bulk fuel supply, which totals over 5 billion gallons. Jet and naval distillate fuel purchased for the Army, Navy, Marine Corps, and Air Force can be supplied by biofuel producers capable of meeting military specifications and cost-competitive with conventional fuels.[33]

Similar to the nation's military, the U.S. airline industry has a large demand for jet fuel—almost 22 billion gallons annually. In 2014, as part of an effort to meet the demand of renewable aviation fuels, DOE, USDA and FAA committed to Farm to Fly 2.0. The federal government, working closely with the aviation industry, will endeavor to enable commercially viable, sustainable renewable jet fuel and necessary supply chains in the U.S. in order to support the goal of one billion gallons jet fuel from biomass for use by 2018.[34]

With Farm to Fly 2.0 as a model, the DoN and USDA are working together in another partnership, Farm-to-Fleet, to ensure markets for renewable biofuel blends as part of the regular, operational fuel purchase and use by the military. $161.33 million of USDA Commodity Credit Corporation (CCC) funds are made available to defray premiums (up to a limit) associated with USDA-approved, US-grown feedstocks for the acquisition of naval distillate or jet fuel. An award was made with the help of CCC funds in October 2015 to AltAir Fuels in Paramount, CA for 77.66 million gallons of F-76 naval distillate fuels consisting of 10% drop-in biofuels made using Midwestern-grown waste beef fats as a feedstock. Farm to Fly 2.0 will continue to seek awards in the domestic bulk fuel supply awards.

Developing a Sustainable Future

DOE and USDA are actively involved in the Sustainability and Greenhouse Gas Accounting working groups of the Global Bioenergy Partnership. The Global Bioenergy Partnership (GBEP) was established in 2006 by a mandate from the G8 + 5 in 2005 and is the preferred venue for USG to talk about sustainability of biofuels internationally. Partners currently include 23 countries and 7 United Nations organizations, with Brazil and Italy as co-chairs. Significant achievements include a GHG Methodological Framework (United States co-chaired with United Nations Foundation) and continued work on development of Sustainability Criteria and Indicators (chaired by United Kingdom). Partners: DOE, USDA, EPA, USTR, and Commerce collaborate with State Department, the National Security Council, and the National Economic Council on a U.S. position going into each of the quarterly GBEP meetings.[35]

---

32 For more nformat on about the DOE-DoN-USDA partnersh p p ease v s t: http://www.energy.d a.m /Pages/DefenseProduct onAct.aspx

33 For more nformat on on Farm to F y 2.0 p ease v s t: http://energy.gov/eere/b oenergy/art c es/farm-fly-20-energy-department-jo ns- n t at ve-br ng-b ofue s-sk es

34 For more nformat on on Farm-to-F eet v s t: http://www.usda.gov/wps/porta /usda/usdamed afb?content d 2013/12/0237.xm &pr ntab e true&content don y true

35 For more nformat on on the G oba B oenergy Partnersh p p ease v s t: http://www.g oba b oenergy.org/

This page is intentionally left blank

www.ingramcontent.com/pod-product-compliance
Lightning Source LLC
Chambersburg PA
CBHW080557190526

45169CB00007B/2806